STUDIES ON
BIO-MEDICAL ANTHROPOLOGY
PROFILES OF HEALTH CULTURES

STUDIES ON

BIO-MEDICAL ANTHROPOLOGY

PROFILES OF HEALTH CULTURES

Edited by

R.K. Mutatkar • Ajit K. Danda • Vikas Bhatt

Indian National Confederation and
Academy of Anthropologists
Jhargram

Indira Gandhi Rashtriya Manav Sangrahalaya
Bhopal

Aryan Books International
New Delhi

STUDIES ON BIO-MEDICAL ANTHROPOLOGY

ISBN: 978-81-7305-372-6

First Published in **2010** by:

INDIAN NATIONAL CONFEDERATION AND ACADEMY OF ANTHROPOLOGISTS
225, Kadam Kanon, Jhargram - 721 507, West Bengal

INDIRA GANDHI RASHTRIYA MANAV SANGRAHALAYA
Shamla Hills, Bhopal - 462 013 (M.P.)

and

ARYAN BOOKS INTERNATIONAL
Pooja Apartments, 4B, Ansari Road, New Delhi-110 002 (India)
Tel.: 23287589, 23255799; Fax: 91-11-23270385
E-mail: aryanbooks@vsnl.com, aryanbooks@gmail.com

Computer Typeset and Printed in India at
ABI Prints & Publishing Co., New Delhi

PREFACE

Indian National Confederation and Academy of Anthropologists (INCAA), the apex professional organization of anthropologists in India organized, in collaboration with its affiliated institution: The Maharashtra Association of Anthropological Sciences (MAAS), an academic discourse: Indian Anthropological Congress at Pune, during February 21 through 23, 2007. The Focal Theme for the Congress was "Human Development: Evolution and Vision". Altogether 222 registered delegates as well as a number of guests and special guests attended the programme that had three Plenary Sessions, two Special Sessions, 16 Scientific Sessions, and four Poster Sessions.

Since the very inception of the INCAA in 2004, the Indira Gandhi Rashtriya Manav Sangrahalaya (IGRMS), Bhopal, stood by INCAA as an extraordinary support. Besides, it has also extended collaboration in bringing out some of the proceedings of the INCAA's academic programme. Studies on Bio-Medical Anthropology, which is serialized as number: 8 INCAA OCCASIONAL PUBLICATIONS, is an illustration of such collaboration.

The volume incorporates a selection of papers related to Biological Anthropology, Bio-Medical Anthropology, and Medical Anthropology; most of which have been presented and discussed at the Indian Anthropological Congress held at Pune.

Professor R.K. Mutatkar, President of the Maharashtra Association of Anthropological Sciences and the Chairman of the Organizing Committee of the Indian Anthropological Congress, Pune, has been kind enough to write a post-script for the volume: On Medical Anthropology.

Ajit K. Danda
Member-Secretary, INCAA

CONTENTS

Set – B

BIO-MEDICAL ANTHROPOLOGY

Set – C

MEDICAL ANTHROPOLOGY

ACKNOWLEDGEMENTS

The Editors of *Studies on Bio-Medical Anthropology* record with appreciation the support received from the Maharashtra Association of Anthropological Sciences (MAAS), Pune; Indira Gandhi Rashtriya Manav Sangrahalaya, Bhopal; Anthropological Survey of India, Kolkata; Tata Institute of Social Sciences, Mumbai; Deccan College Post-Graduate and Research Institute, Pune; Indian Association for the Study of Traditional Medicine, Mumbai; University of Poona, Pune; toward organization of the Second Indian Anthropology Congress, of which the present volume is part of the academic outcome. The young scholars from the MAAS and Department of Anthropology, University of Pune deserve special appreciation not only for providing logistic support but also academic organization, such as classifying abstracts, composing relevant sessions and classifying research papers, under the guidance of Prof. Mutatkar, Vice Chairman of INCAA at Pune. Saju Joseph, Dr. Abhay Kudale, Dr. Hemant Apte, Dr. R.D. Gambhir and Shantanu Ozarkar discharged the responsibilities with great academic commitment. Special thanks are due to the Director, Indira Gandhi Rashtriya Manav Sangrahalaya, Bhopal; and its officers responsible for Publications for their interest in, and initiative toward, bringing out the volume quickly. The Editors also acknowledge with thanks the unalloyed cooperation and support received from the contributors to this volume. Shri Vikas Arya of Aryan Books International, New Delhi has made the Editors indebted through his initiative in bringing out the *Studies on Bio-Medical Anthropology* in minimum possible time.

— Editors

INTRODUCTION

I

The Indian National Confederation and Academy of Anthropologists (INCAA), under the auspices of the University of Poona, organized the Second Indian Anthropological Congress during February 21-23, 2007, as part of its overall academic commitments. The Maharashtra Association of Anthropological Sciences (MAAS), one of the affiliated institutions of INCAA, undertook the responsibility of coordinating the entire programme. Indira Gandhi Rashtriya Manav Sangrahalaya, Bhopal; Anthropological Survey of India, Kolkata; Tata Institute of Social Sciences, Mumbai; Deccan College Post-Graduate and Research Institute, Pune; and the Indian Association for the Study of Traditional Asian Medicine, Mumbai extended their generous collaboration to this endeavour.

Altogether 222 registered delegates as well as a number of guests and special guests attended the programme that included, besides the Inaugural Session, as envisaged by INCAA and MAAS, three Plenary Sessions, two Special Sessions, 16 Scientific Sessions and four Poster Sessions.

Altogether 54 research papers belonging to the Biological Anthropology, Social/Cultural Anthropology, and Medical Anthropology were presented and discussed in the Indian Anthropological Congress. Among them, 11 papers dealt with issues exclusively related to Biological Anthropology. The number of contributions on Social/Cultural Anthropology was 21 and the other 22 contributions belong to Medical Anthropology or a combination of both Biological Anthropology and Medical Anthropology.

Ordinarily as it happens in such cases, our experience with the contributions received for the Indian Anthropological Congress was no different in any way. On scrutiny of the papers that were presented and discussed in various scientific sessions, we found them greatly vary in terms of overall quality, although the bulk of them were very much rich so far as the data are concerned. Papers on Biological Anthropology and Medical Anthropology are particularly rich in this respect, despite the fact that the majority of such contributions are exploratory in nature. They, nevertheless, are also rich in analysis, though several among them demand further exercise toward appropriate interpretation. Contributions to Social/Cultural Anthropology stand out rather distinct in such characters and on several other counts. They would, therefore, be dealt with separately.

II

In view of certain practical considerations, it has also been decided to bring out the papers related to Biological Anthropology and Medical Anthropology together. Thus, out of 33 contributions received for deliberations in the eight Bio-Medical related Scientific Sessions of the Indian Anthropological Congress, 28 papers have been chosen for inclusion in the present volume. Besides, Prof. R.K. Mutatkar contributed an article on Medical Anthropology for this volume, though it was not presented or discussed at the Congress.

For operational advantage they have, however, been organized into three sets: A through C. Set-A or the first set of eight papers of this volume more or less fits into the scheme of Biological Anthropology. The second set or Set-B also has eight papers, that although primarily belong to Biological Anthropology, have close bearing upon Medical Anthropology. They, in a way, represent an assumed intermediate phase touching upon both Biological Anthropology and Medical Anthropology. Set-C includes 12 papers, all qualitatively belonging to Medical Anthropology. The contributions included in this volume have, therefore, been arranged accordingly.

This speaks rather of one-half of the entire story, since for a wholesome account of our academic experiences gained through exposures to the Second Indian Anthropological Congress one needs to pay equally close attention to the contributions belonging to Social/Cultural Anthropology as well. As already indicated, we propose to bring out a separate volume including the papers on Social/Cultural Anthropology. Until that has been achieved, let us refrain from making any generalized statements on the basis of contributions included in the present volume. We can, nevertheless, add that the major thrusts of each of the individual contributions when looked at collectively have been found sufficiently meaningful in the context of commitments and professed objectives of INCAA.

III

Anthropology as the only holistic science of man, studies human variation, human nature, and human creations as evolved historically, blending the biological and cultural facts and perspectives. The holistic perspectives of mankind during all periods of time encompassing the dynamic triangular relationship of man-environment-supernatural is the heritage of anthropology, which anchors the civil society for peace and harmony.

Anthropology contributed to the body of knowledge by studying the cultural and biological variability which got enshrined in the U.N. Charter of Human Rights respecting the principles of cultural relativism. Results of anthropological studies also led to the abolition of racial discrimination. It is, nevertheless, yet not recognized that human genetics, ethnography, and qualitative research methods and many more such approaches, concepts, tools and techniques are the gifts of anthropology that are being used and marketed by other disciplines.

The post-colonial era has shifted the focus of humanity to egalitarian order, poverty alleviation, and development of issues encompassing health and education. Human development is the key concept used for measurement, as well as an approach which is essentially anthropological in nature, since it signals comprehensive and sustainable development flowing from the theories of culture. New dimensions have been added to the objective due to desegregation of culture and lifestyle, and due to global effects of technology.

IV

Indian anthropology is broadly visible in the University Departments, Tribal Research Institutes, Anthropological Survey of India, Research Institutes of ICMR, ICSSR, etc., committees and commissions dealing with backward communities, and in the NGO Sector undertaking Health and Development Action Research. In response to the demand of the local requirements, anthropologists organized themselves to constitute academic bodies. In reality, INCAA has shaped up itself as a national platform to confederate networking of such regional bodies. As an academy it, nevertheless, has a vision to address national issues without interfering or jeopardizing the interests of the local outfits. In fact, the regional associations are expected to gain strength from national linkage and by participating in the national endeavours. The mutuality between the INCAA and MAAS demonstrates the same with evidence.

While anthropology is the only discipline which primarily focuses on people, describing the cultural, social and economic framework that leads to poverty and deprivation, including de-habilitation, due primarily to development of 'mainstream', it has not yet addressed the issues at the policy level. INCAA, therefore, has the responsibility to play the role of the facilitator at the micro as well as at the macro policy planning level by using the network of anthropologists and strengthening the professional database.

When other academic disciplines, including bio-medical applied disciplines, from their attention to the outcomes, labeling them as problems, use curative symptomatic methods to solve them, anthropology, being equipped in in-field strategies opt for studying the processes in-depth in order to provide preventive solutions. This gets reflected even in the academic exercises that are being organized for some kind of assessment of anthropologists' performance record. The empirical study-accounts deliberated and dealt with in the anthropological conferences, including the Indian Anthropological Congress, amply justify this position.

R.K. Mutatkar
Ajit K. Danda
Vikas Bhatt

1

ANTHROPOMETRIC STUDY AMONG THE JENU KURUBAS OF COORG

C.E. Subramanya, K. Rajasekhara Reddy
and K. Bhamini Raghavaiah

INTRODUCTION

Anthropometry represents the typical and traditional tool of Physical Anthropology. Anthropometric characters, in spite of their interaction with environmental forces and polygenic nature, studied under common nutritional, ecological and social factors, are said to show population variation which is the combined effect of natural selection and genetic drift.

Most of the physical anthropological researches (such as demography, anthropometry, etc.) among various communities are still in their infancy. Even such studies are lacking in number as far as the populations of Coorg District in South Karnataka are concerned. Hence, the anthropometric characters among the Jenu Kuruba population have been undertaken for the present study.

The Jenu Kurubas, a large tribal population, are confined to the forests of Coorg District. Their main occupation is honey collection. They are also engaged as labourers in coffee plantations in the district. They are medium-statured people, living in thatched huts or houses, built of mud. Some live in line houses provided by the owners of coffee plantations. They are non-vegetarians and depend on forest produce which they collect through hunting and gathering.

Aims and Objectives of the Present Study

1. To find out the anthropometrical profile of Jenu Kuruba population,
2. To understand the variation in biological composition among the Jenu Kuruba population in respect of anthropometric aspects, and

3. To find out possible anthropometric indices in order to know the racial characteristics of the population.

MATERIAL AND METHODS

A sample of 400 Jenu Kuruba individuals of either sex (males: 200, Females: 200) from Coorg District of Karnataka in the age-group of 20-50 years has been included in the present study. Seventeen measurements are taken, out of which eleven are on head and face, and six are on body form for the present study. On the basis of these measurements, indices are calculated. The measurements are recorded according to methods and techniques of Martin and Saller (1957), and Singh and Bhasin (1989).

RESULTS AND DISCUSSION

Anthropometric Measurements

Stature: The Jenu Kuruba male stature ranges from 138.4 cm to 173.4 cm with a mean value of 152.25 ± 1.20, while the female stature ranges from 132.4 cm to 164.0 2 cm with a mean value of 144.85 ± 0.13 cm. The stature classification reveals that the majority of the Jenu Kuruba males come under upper short category (34.0%) and their female counterparts fall under short category (36.5%).

Sitting Height: The Jenu Kuruba male and female sitting height vertex values range from 64.0 cm to 87.6 cm with a mean value of 78.27 ± 0.15 cm and from 62.0 cm to 75.0 cm with a mean value of 71.87 ± 0.10 cm, respectively. The majority of the males and females come under low category (50.0% and 49.0% respectively).

Body Weight: The Jenu Kuruba male's body weight values range from 31.0 kg to 62.0 kg with a mean value of 46.65 ± 0.60 kg whereas the female population ranges from 28.0 kg to 52.0 kg with an average value of 40.05 ± 0.05 kg. The majority of the male (35.0%) and female (34.0%) individuals fall under the range of 40.0 to 44.9 kg.

Head Length: The Jenu Kuruba male's head length values range from 17.3 cm to 19.9 cm with a mean value of 18.51 ± 0.04 cm and the female's head length ranges from 16.1 cm to 19.9 cm with an average value of 17.32 ± 0.05 cm. A relatively higher proportion of the Jenu Kuruba males (59.50) and females (49.50%) fall under the medium category.

Head Breadth: The head breadth values of males range from 12.1 cm to 14.9 cm with a mean value of 13.71 ± 0.03 cm and those of females range from 11.9 cm to 14.8 cm, with an average value of 13.23 ± 0.01 cm. The majority of male and female individuals fall under a very narrow category (84.0 and 94.0 respectively).

Head Circumference: The head circumference values range from 50.0 cm to 58.0 cm with a mean value of 54.19 ± 0.13 cm among the males, whereas these range from 49.4 cm to 59.0 cm with a mean value of 53.01 ± 0.12 cm among the females. The majority of the male individuals (55.50%) come under the range of 54.0 cm to 55.9 cm, while most of the female individuals (51.0%) fall under the range of 52.0 cm to 53.9 cm.

Minimum Frontal Breadth: The Jenu Kuruba male's minimum frontal breadth values range from 9.8 cm to 11.8 cm with an average value of 10.87 ± 0.02 cm and its female minimum frontal breadth ranges from 9.6 cm to 12.9 cm, with an average value of 11.21 ± 0.08 cm. The majority of the male individuals come under the above broad type (49.50%), whereas the female individuals come under medium type (59.50%).

Bizygomatic Breadth: The Jenu Kuruba male bizygomatic breadth values range from 10.2 cm to 13.4 cm with a mean value of 14.40 ± 0.03 cm and female bizygomatic breadth ranges from 11.4 cm to 12.8 cm with a mean value of 11.82 ± 0.03 cm. A large number of both male (31.0%) and female (68.0%) populations of Jenu Kuruba belong to very narrow type.

Physiognomic Facial Height: The male Jenu Kuruba physiognomic facial height ranges from 14.9 cm to 18.7 cm with a mean value of 16.69 ± 0.07 cm and the female's physiognomic facial height ranges from 14.2 cm to 18.4 cm with a mean value of 15.91 ± 0.06 cm. The majority of the males (43.50%) and females (44.50%) fall under the range of 16.0 cm to 16.9 cm and 15.1 cm to 16.0 cm, respectively.

Physiognomic Upper Facial Height: The physiognomic upper facial height values in the male series range from 10.8 cm to 14.6 cm with an average value of 12.75 ± 0.06 cm and in the female series, it ranges from 10.2 cm to 15.0 cm with a mean value of 11.98 ± 0.06 cm. The majority of the males (43.00%) come under the range of 12.0 cm to 12.9 cm whereas a high proportion of females (44.50%) fall under the range of 11.0 cm to 11.9 cm.

Morphological Facial Height: The morphological facial height among males and females of Jenu Kuruba varies from 9.8 cm to 12.1 cm with a mean value of 11.17 ± 0.02 cm and 8.8 to 11.4 cm with a mean value of 11.17 ± 0.02 cm respectively. The majority of Jenu Kuruba males (59.52%) and females (67.50%) fall under very low type.

Morphological Upper Facial Height: The morphological upper facial height of Jenu Kuruba males ranges from 5.7 cm to 7.8 cm with the mean value of 6.50 ± 0.03 cm and that of females ranges from 5.3 cm to 7.0 cm with a mean value of 5.85 ± 0.02 cm. A high proportion of males (39.0%)

and females (42.0%) fall under the range of 6.5 cm to 6.9 cm and 5.6 cm to 6.0 cm respectively.

Nasal Height: The nasal height of Jenu Kuruba males and females varies from 3.9 cm to 5.7 cm with a mean value of 4.54 ± 0.02 cm and 3.6 cm to 5.3 cm with a mean value of 4.13 ± 0.02 cm respectively. A large proportion of males (48.50%) and females (49.0%) fall under below medium and short category respectively.

Nasal Breadth: The nasal breadth among Jenu Kuruba males ranges from 3.3 cm to 4.6 cm with a mean value of 3.73 ± 0.07 cm and among females ranges from 2.9 cm to 4.4 cm with an average value of 3.49 ± 0.02 cm. The majority of males (52.0%) and females (43.0%) fall under above medium and medium type respectively.

Biacromial Breadth: The biacromial breadth values of males range from 30.7 cm to 34.5 cm with a mean value of 33.34 ± 0.07 cm and its values of females range from 27.6 cm to 31.9 cm with a mean value of 28.43 ± 0.06 cm. The majority of the Jenu Kuruba males (36.50%) and females (45.0%) come under the range of 34.0 cm to 34.9 cm and 28.0 cm to 28.9 cm respectively.

Upper Arm Circumference: The Jenu Kuruba males' upper arm circumference ranges from 18.8 cm to 26.2 cm with a mean value of 23.14 ± 0.11 cm and its values among the females range from 17.8 cm to 25.9 cm with a mean value of 21.08 ± 0.13 cm. Fairly high proportions of males (46.50%) and females (44.50%) come under the ranges of 21.0 cm to 22.9 cm and 20.0 cm to 21.9 cm respectively.

Chest Circumference: The Jenu Kuruba males' chest circumference ranges from 71.8 cm to 88.9 cm with a mean value of 79.30 ± 0.27 cm while its female chest circumference ranges from 58.6 cm to 81.2 cm with a mean value of 67.22 ± 0.42 cm. A large proportion of males (43.50%) fall under the range of 78.0 cm to 81.9 cm whereas a high proportion of females (38.0%) comes under the range of 65.0 cm to 69.9 cm.

Indices

Cephalic Index: The cephalic index range values from 66.9 to 84.3 units with a mean value of 73.35 ± 0.30 among the males of Jenu Kuruba, revealing that the majority of the males belong to dolicocephalic (47.50%) category. The values of the females range from 67.9 to 83.1 units with an average value of 73.07 ± 0.27, showing that a higher proportion of the females (45.50%) comes under hyper-dolicocephalic category.

Nasal Index: The nasal index values among the males range from 48.9 to 10.10 units with an average value of 71.35 ± 0.93 and among the females, they range from 49.9 to 100.1 units with a mean value of 71.57 ± 0.94. Majority of males (43.50%) and females (39.50%) fall under leptorhinae type.

Physiognomic Facial Index: The physiognomic facial index ranges from 116.2 to 141.2 units with an average value of 128.22 ± 0.35 among the males and from 127.0 to 152.3 units with a mean value of 140.47 ± 0.43 among the females. A large proportion of males (38.00%) come under the range of 125.0 to 129.9 whereas among females, 30.0% falls under the range of 136.0 to 140.9.

Physiognomic Upper Facial Index: The physiognomic upper facial index ranges from 65.2 to 80.2 units with a mean value of 73.70 ± 0.29 in the male series and from 71.0 to 81.2 units with an average value of 77.17 ± 0.19 in the female series. A high proportion of males (43.50%) falls under the range of 70.0 to 74.9 and that of females (39.50%) comes under the range of 78.0 to 80.9.

Morphological Facial Index: The morphological facial index varies from 75.2 to 96.4 units with a mean value of 86.10 ± 0.44 among the males and from 74.9 to 93.2 units with an average value of 84.28 ± 0.31 among the females. The large proportion of males (28.50%) falls under hyper mesoprosopic category whereas a higher proportion of females (34.50%) comes under leptoprosopic category.

Morphological Upper Facial Index: The morphological upper facial index values range from 39.3 to 58.6 units with an average value of 49.12 ± 0.41 among the males and from 42.2 to 59.7 with an average value of 51.85 ± 0.32 among the females. A large proportion of males (29.50%) falls under euryen type while a higher proportion of females (38.50%) comes under mesen type.

Sagittal-Naso Facial Index: The sagittal-naso facial index value ranges from 39.7 to 47.1 units with an average value of 43.02 ± 0.13 among the males and from 40.9 to 46.4 units with an average value of 43.72 ± 0.10 among the females. A high proportion of males (34.50%) and females (46.50%) falls under the range of 42.0 to 43.9 and 43.0 to 44.9 respectively.

Biacromial Breadth Stature Index: The biacromial breadth stature index ranges from 20.8 to 23.9 units with a mean value of 22.66 ± 0.05 among the males and from 19.6 to 23.2 unit with a mean value of 21.70 ± 0.04 among the females. A large proportion of the males (45.50%) comes under medium shoulder category whereas a higher proportion of females (47.0%) falls under medium shoulder category.

Chest Girth Stature Index: The chest girth stature index of males ranges from 47.2 to 60.3 units with a mean value of 53.70 ± 0.25 units and that of females varies from 44.2 to 58.9 units with an average value of 52.82 ± 0.23 among the Jenu Kuruba. This reveals that a high proportion of males (48.0%) and the majority of females (54.50%) falls under medium chest category.

Robusticity Index: The robusticity index ranges from 6.4 to 38.2 units with a mean value of 22.32 + 0.48 among the males and from 9.0 to 37.2 units with an average value of 23.80 + 0.47 among the females. A large proportion of males (33.0%) comes under medium category whereas a relatively lower proportion of females (28.50%) falls under good category.

It is evident from the above figures that the majority of the males and females among the Jenu Kurubas belong to medium stature, medium robust body, dolico-cephalic head, leptorhinae nose, mesoprosopic to leptoprosopic face, medium shoulder and medium chest type.

BIBLIOGRAPHY

Martin, R. and K. Saller. 957. *Leherbuen der Anthropologic*. Vol. 1 and 11. Stuttgart: Gutsav Fischer Veslag.

Singh, I.P. and M.K. Bhasin. 1989. *A Laboratory Manual on Biological Anthropology*. New Delhi: Kamal Raj Enterprises.

2

QUANTITATIVE ASSESSMENT OF SELECT DERMATOGLYPHIC FEATURES AMONG THE PEOPLE OF BHIMTAL (DISTRICT NAINITAL, UTTARANCHAL)

Purnima Parashar and C.J. Edwin

INTRODUCTION

The quantitative method in dermatoglyphics is relatively a new approach developed to study dermal ridges. This method enables the application of significance tests as also the elimination of arbitrariness of certain qualitative classification (Reddy, 2006). The widely studied quantitative dermatoglyphics characters include Total Finger Ridge Count (TFRC), Absolute Finger Ridge Count (AFRC), Pattern Intensity Index, Dankmeijers Index, Furuhata's Index on fingers and interdigital ridge counts (a-b, b-c, c-d) and ATD angle on palms. Holt (1968) opines that of all the quantitative characters, TFRC is strongly determined by heredity excepting only 5% of its total variability caused by environmental effects. Jantz and Hawkinson (1979) believe that in order to discern meaningful population patterning, individual ridge counts provide additional information. While, similar work was done by Sunderland and Coope (1969) on Jivaro of Ecuador. Harvey and Suter (1983) discerned evidences for a major gene effect in the distribution of digital ridge counts. Harich *et al.* (2000) studied Berbers from Morocco and showed that bilateral symmetry was more pronounced than sexual differences in overall distribution, whereas they exhibit high value of Total Finger Ridge Count (TFRC).

The Absolute Finger Ridge Count (AFRC) also has been utilized to a greater extent in the biological comparison of populations (Mukherjee, 1967; Chakraborty and Malhotra, 1981; Karmakar *et al.*, 2005). Owing to strong correlation with Total Finger Ridge Count (TFRC), the Pattern Intensity

Index (PII) may have an important genetic component (Salzano and Benevides, 1979; Karmakar *et al.*, 2005). Newman (1960) considered the indices of Pattern Intensity as one of the important criteria for the evaluation of the biologically meaningful differences between the groups of populations. The amount of variation in individual pattern frequencies (Pattern Intensity Index, Dankmeijer's Index, and Furuhatas's Index) influences the variation in quantitative characters and signifies the sexual dimorphism (Narahari and Padmaja, 2006).

Palmar Quantitative Character, Interdigital Ridge Count and ATD Angle are studied for understanding the mode of inheritance and for delineating the bisexual and inter- and intra-population variation (Mahapatra and Behara, 1970; Pertia, 1974; Mate, 1975; Florence *et al.*, 1975; Plato *et al.*, 1975; Prokopec and Sevidy, 1976; Murillo *et al.*, 1977; Harvey and Suter, 1983; Reddy *et al.*, 2000; Igbigbi and Masamati, 2005; Reddy, 2006).

Where some clinical based quantitative dermatoglyphics work was also done by some authors (Okajima *et al.*, 1984; Davee *et al.*, 1989; Kamali and Mavalwala, 1990; Frazer, 1994; Kohli and Bhalla, 2000; Basu *et al.*, 2005; Inamdar *et al.*, 2006; Natekar *et al.*, 2006; Banik and Mukherjee, 2006).

LAND AND PEPOLE

Data for the present study were collected from Bhimtal Block, District Nainital, Uttaranchal. This falls in the Kumaon region of the state. Bhimtal is located in the northern part of district Nainital and its about 22 km from Nainital. It is located at an altitude of 1,371 metres above sea level. It lies between 78° 80' and 80° 14' East Longitude and 29° 00' and 29° 05' North Latitude. The place derives its name from a legend related to 'Bhim', one of the Pandavas from the epic *Mahabharata*. The famous Bhimtal Lake is situated in the centre of Bhimtal town and is one of the largest lakes in the district.

The most dominant people numerically as well as economically are Brahmins and Thakurs. The Aryas (Schedule Caste) also form a significant proportion in the population. All the three communities were largely agricultural in occupation. The populations were sessile and have been living there for generations (Table No. 1).

MATERIAL AND METHODS

The data for the present study comprises bilateral inked fingers and palmar prints of 174 males and 183 females of Bhimtal Block, district Nainital, Uttaranchal. Ad-hoc sampling method was used. The age of the individuals ranges from 10 to 50 years. The analysis of finger and palm prints have

Table No. 1
Chief Statistics of Bhimtal Block

Total Area	9795.37 Hectare
Total Population: Persons Males **Females**	37,539 19,747 17,792
Scheduled Caste (S.C)	9,420
Scheduled Tribe (S.T.)	40
Literacy Rate: **Absolute** **Percentage**	2,3361 62.23
Total Cultivated Area	727 sq. km.
Agricultural Area	9268 sq. km.

been according to Cummins and Midlo (1961). Chi-square and Z-test were computed to know the significance of bimanual, bisexual and intercommunity differences.

The following quantitative dermatoglyphics features were observed:

(A) On Finger Balls:

1. Total Finger Ridge Count (TFRC)
2. Absolute Finger Ridge Count (AFRC)
3. Indices:
   (a) Pattern Intensity Index
   (b) Dankmeijer's Index
   (c) Furuhatas's Index

(B) On Palms:

4. Palmar Interdigital Ridge Counts
5. ATD Angle.

OBSERVATIONS

1. Total Finger Ridge Count (TFRC)

Total Finger Ridge Count (TFRC) for right-left handed males and females of all the three communities, viz., Aryas, Brahmins and Thakurs are shown in Table No. 2. Thakurs showed higher mean values for both right (65.32) and left (66.03) in males as well as in females, right (57.96) and left (56.35) hands. The comparison of the means show no right-left differences.

Table No. 2
Total Finger Ridge Count: Male-Female, Right-Left Comparison

	MALES						FEMALES					
	Aryas		Brahmins		Thakurs		Aryas		Brahmins		Thakurs	
	R	L	R	L	R	L	R	L	R	L	R	L
N	58	58	57	57	53	53	64	64	56	56	52	52
Range	23-99	22-99	13-115	10-149	17-100	22-105	7-116	10-91	12-80	6-78	18-93	21-23
Mean	57.83	60.34	56.40	62.59	65.32	66.03	57.19	56.28	53.55	52.43	57.96	56.35
S.D.	19.93	62.95	24.50	21.82	17.76	19.80	19.50	19.56	17.26	19.47	18.86	19.80
C.V.	34.46	66.03	43.45	34.66	27.19	29.98	34.09	34.75	32.33	37.13	32.54	35.14

However, the male-female comparisons show (Table No. 3) significant differences in the Brahmins (at 1 per cent) and among Thakurs (at 5 per cent). The Aryas do not show any bisexual differences. When both the sexes were pooled, we did not get any intercommunity differences.

2. Absolute Finger Ridge Count (AFRC)

Absolute Finger Ridge Count (AFRC) for right-left hands of males and females of all the three communities are shown in Table No. 4. Here also, the Thakurs showed higher mean values for both right (91.40) and left (87.26) in males as well as females' right (78.58) and left (77.12) hands. Comparing the mean values, no significant differences were seen for right-left hands. Male-female differences are significant (at 5 per cent) for Brahmins only. When both sexes were pooled and inter-community comparisons were made, significant differences (at 5 per cent) were seen only between Brahmins and Thakurs (Table No. 5).

3. Indices

The Pattern Intensity Index (PII) of males and females is shown in Table No. 6. The mean value varies from 12.14 to 13.89. There are neither any significant male-female nor intercommunity differences in mean values.

The mean value for Dankmeijer's Index and Furuhata's Index (Table Nos. 7 & 8) vary from 8.00 to 24.03 and 46.36 to 81.43 respectively. There are no significant right-left and male-female differences in the case of these two indices. There are no inter-community differences. Although the Dankmeijer's Index is higher in females in case of both hands, Furuhata's Index does not show any such regular pattern in variation.

Table No. 3

Total Finger Ridge Count: Right and Left Hands Combined for Males, Females and Both Sexes Pooled

	MALES			FEMALES			BOTH SEXES POOLED		
	Aryas	**Brahmins**	**Thakur**	**Aryas**	**Brahmins**	**Thakurs**	**Aryas**	**Brahmins**	**Thakurs**
N	58	57	53	64	56	52	122	113	105
Range	38-198	31-248	39-203	17-183	22-158	34-169	17-198	22-248	34-203
Mean	120.69	124.05	132.32	113.76	107.01	115.78	117.07	115.69	123.84
S.D.	35.92	35.79	35.08	36.61	33.25	36.81	36.33	35.22	36.64
C.V.	29.81	28.85	26.51	32.19	31.06	31.69	36.64	30.45	29.58

MALE-FEMALE COMPARISON
ARYAS: Z= 2.36 Significant at 5 per cent level
BRAHMINS: Z= 2.62 Significant at 1 per cent level
THAKURS: Z= 1.06 Not Significant

Table No. 5

Absolute Finger Ridge Count: Right and Left Hands Combined for Males, Females and Both Sexes Pooled

	MALES			FEMALES			BOTH SEXES POOLED		
	Aryas	**Brahmins**	**Thakur**	**Aryas**	**Brahmins**	**Thakurs**	**Aryas**	**Brahmins**	**Thakurs**
N	58	57	53	64	56	52	122	113	105
Range	45-340	31-348	34-333	17-298	22-295	42-319	17-340	22-348	34-333
Mean	168.86	161.72	180.15	148.03	135.61	158.36	158.02	146.92	169.44
S.D.	75.27	65.63	70.16	65.47	55.98	73.69	70.82	62.58	72.39
C.V.	44.58	40.58	38.94	44.23	41.27	46.48	44.82	42.59	42.72

MALE-FEMALE COMPARISON
ARYAS: Z= 1.62 Not Significant
BRAHMINS: Z= 2.27 Significant at 5 per cent level
THAKURS: Z= 1.54 Not Significant

INTER COMMUNITY COMPARISON
ARYAS-BRAHMINS: Z= 1.27 Not Significant
ARYAS-THAKURS: Z= 1.20 Not Significant
BRAHMINS-THAKURS: Z= 2.25 Significant at 5 per cent level

Table No. 4

Absolute Finger Ridge Count: Male-Female, Right-left Comparison

	MALES						FEMALES					
	Aryas		Brahmins		Thakurs		Aryas		Brahmins		Thakurs	
	R	L	R	L	R	L	R	L	R	L	R	L
N	58	58	57	57	53	53	64	64	56	56	52	52
Range	22-171	15-171	13-171	10-179	27-160	22-174	7-115	16-136	12-155	6-140	18-170	17-132
Mean	84.96	82.22	82.72	77.35	91.40	87.26	74.68	70.28	67.50	68.14	78.58	77.12
S.D.	37.26	36.95	34.31	35.23	37.67	36.64	34.57	34.34	28.98	29.74	40.41	34.84
C.V.	43.85	44.94	41.48	45.45	41.21	41.98	46.30	48.85	42.93	43.64	51.43	45.18

Table No. 6

Distribution of Pattern Intensity Index

	MALES			FEMALES		
	Aryas	Brahmins	Thakurs	Aryas	Brahmins	Thakurs
N	58	57	53	64	56	52
Mean	13.89	13.07	13.83	12.69	12.14	13.40
S.D.	3.65	3.32	3.36	3.46	3.68	4.23
C.V.	26.27	25.40	24.29	27.26	30.31	31.56

Table No. 7

Bimanual/Bisexual Distribution of Dankmeijer's Index

	MALES			FEMALES		
	Right	Left	Both Hands	Right	Left	Both Hands
Aryas	3.79	12.71	8.00	21.00	26.85	24.03
Brahmins	7.48	18.18	12.03	17.95	23.86	21.08
Thakurs	7.32	10.78	8.88	14.81	17.64	16.19

Table No. 8

Bimanual/Bisexual Distribution of Furuhata's Index

	MALES			FEMALES		
	Right	Left	Both Hands	Right	Left	Both Hands
Aryas	87.42	75.64	81.43	50.25	59.02	54.44
Brahmins	64.07	48.61	56.03	41.71	51.46	46.36
Thakurs	93.18	68.00	79.78	79.41	72.85	76.08

4. Inter-digitial Ridge Counts

Table No. 9 gives the range, mean, standard deviation of the inter-digital ridge count among males' right and left hand of all the three communities. The mean values are lower in case of b-c ridge count as compared to a-b and c-d ridge counts. There were no significant differences in right-left hand when comparisons were made except the right-left hand comparison among Brahmins showed significant values at 5 per cent.

Table No. 10 gives the inter-digital ridge count among females. Here also, the mean values are lower in case of b-c ridge count as compared to a-b and c-d ridge counts. There were no right-left significant differences. The inter-community comparison for the right and left hand of males and females indicates all non-significant values.

5. Atd Angle

Table No. 11 shows the range, modal value, mean and standard deviation of the ATD angle for males' right and left hands in all the three communities. The modal values are 41° and 47° for right and left hand in Aryas, 45° and 39° for Brahmins and 43° and 39° for Thakurs. The comparison of means showed insignificant value for right-left comparison in all the three communities.

Table No. 12 represents the ATD angle among females. The modal value for Aryas' right and left hand are 46° and 45°, for Brahmins 46° in both hands and 41° and 43° in Thakurs. The comparison of means here also showed non-significant values for right-left comparison. There was no male-female significant value among the Aryas and Brahmins. However, the Thakurs' male-female comparison for right hand indicates differences significant at 5 per cent. The inter-community comparison shows significant differences at 1 per cent for the right hand of Arya-Thakur female only.

DISCUSSION

The overall picture of ridge counts does not show a clear pattern of variation in the Total Finger Ridge Count and Absolute Finger Ridge Count. Looking at the indices derived from finger ball pattern, namely, Pattern Intensity Index, Dankmeijer's Index and Furuhata's Index, one sees no bimanual, bisexual and inter-community differences. However, with regard to Dankmeijer's Index, the females show higher values than the males indicating a trend which is not significant for females to have relatively more arches than whorls as compared with males.

a-b ridge count is the number of ridges occurring in the second inter-digital area between the a and b triradii. It is the best known and the most

Table No. 9

Distribution of Interdigital Ridge Count among Males in all the three communities

	ARYAS						BRAHMINS						THAKURS					
	a-b		b-c		c-d		a-b		b-c		c-d		a-b		b-c		c-d	
	R	L	R	L	R	L	R	L	R	L	R	L	R	L	R	L	R	L
N	58	59	55	54	55	54	60	60	59	56	59	56	54	52	53	46	53	46
Range	21-46	18-54	9-36	14-38	15-50	22-48	22-51	23-51	6-37	6-36	13-46	20-46	22-53	21-53	12-37	17.45	7-50	9-50
Meant ± SEM	39.17 ± 1.05	37.89 ± 0.91	24.81 ± 0.92	25.12 ± 0.75	33.29 ± 1.04	32.87 ± 0.91	38.15 ± 0.72	39.43 ± 0.67	24.83 ± 0.84	23.51 ± 0.79	34.40 ± 0.87	31.80 ± 0.92	38.42 ± 1.10	38.40 ± 1.03	25.15 ± 0.79	25.28 ± 0.80	33.69 ± 1.15	33.63 ± 1.19
S.D± SESD	8.03 ± 0.74	6.99 ± 0.64	6.83 ± 0.65	5.57 ± 0.53	7.71 ± 0.73	6.73 ± 0.64	5.58 ± 0.50	5.25 ± 0.47	6.68 ± 0.59	5.97 ± 0.56	6.72 ± 0.61	6.92 ± 0.65	8.11 ± 0.78	7.48 ± 0.73	5.80 ± 0.56	5.43 ± 0.56	8.43 ± 0.81	8.09 ± 0.84

Table No. 10

Distribution of Interdigital Ridge Count among Females in all the three communities

	ARYAS						BRAHMINS						THAKURS					
	a-b		b-c		c-d		a-b		b-c		c-d		a-b		b-c		c-d	
	R	L	R	L	R	L	R	L	R	L	R	L	R	L	R	L	R	L
N	67	68	62	63	62	63	58	59	56	53	57	53	53	53	48	48	48	48
Range	16-49	20-57	13-37	11-37	20-50	19-47	26-49	26-48	11-37	14-37	11-45	16-51	25-48	26-56	14-36	16.35	19-45	17-40
Meant ± SEM	38.89 ± 0.86	39.79 ± 0.85	24.09 ± 0.69	25.52 ± 0.70	34.17 ± 0.81	33.20 ± 0.86	39.94 ± 0.64	39.11 ± 0.53	25.94 ± 0.75	26.54 ± 0.70	33.31 ± 1.06	34.56 ± 1.07	38.92 ± 0.60	39.58 ± 0.75	25.31 ± 0.75	25.41 ± 0.68	35.20 ± 0.79	33.70 ± 0.64
S.D ± SESD	7.08 ± 0.61	7.08 ± 0.60	5.49 ± 0.49	5.63 ± 0.50	6.44 ± 0.57	6.82 ± 0.60	4.93 ± 0.45	4.14 ± 0.38	5.63 ± 0.53	5.12 ± 0.49	8.05 ± 0.75	7.86 ± 0.76	4.39 ± 0.42	5.48 ± 0.53	5.23 ± 0.53	4.71 ± 0.48	5.50 ± 0.56	6.31 ± 0.64

Table No. 11
Variation in ATD angle in three communities among Males

	ARYAS		BRAHMINS		THAKURS	
	Right	**Left**	**Right**	**Left**	**Right**	**Left**
N	58	59	60	60	54	53
Range	29°-62°	25°-85°	24°-87°	31°-70°	26°-78°	29°-67°
Modal value	41°	47°	45°	39°	43°	39°
Mean ± SEM	42.30 ± 0.89	42.50 ± 1.21	42.20 ± 1.12	41.31 ± 0.90	42.02 ± 1.18	41.15 ± 1.12
S.D. ± SESD	6.81 ± 0.63	9.34 ± 0.86	8.67 ± 0.79	6.99 ± 0.63	8.67 ± 0.83	8.18 ± 0.79

satisfactory ridge count on the palm, partly because there is usually no pattern between the triradii. Like other ridge counts, it is also independent of age (David, 1984). There are no significant bimanual differences. The correlation between the two hands is high, the a-b ridge count being very slightly higher on left hand except in Arya and Thakur males, where the right hand shows higher a-b ridge count. The a-b ridge count does not show any sex differences (Holt, 1968). The inter-community comparison does not indicate any significant differences in the present study.

It should be remembered that the ATD angle varies with age. In all the communities observed, both hands show high correlation, there being no bisexual differences either. The inter-community differences in ATD angle

Table No. 12
Variation in ATD angle in three communities among Females

	ARYAS		BRAHMINS		THAKURS	
	Right	**Left**	**Right**	**Left**	**Right**	**Left**
N	68	57	59	59	53	54
Range	23°-56°	29°-55°	27°-62°	27°-70°	32°-60°	31°-70°
Modal value	46°	45°	46°	46°	41°	43°
Mean ± SEM	41.85 ± 0.97	41.95 ± 1.17	43.85 ± 0.83	44.23 ± 0.88	45.40 ± 0.96	44.61 ± 0.95
S.D. ± SESD	8.01 ± 0.68	8.84 ± 0.82	6.40 ± 0.58	6.77 ± 0.62	7.06 ± 0.68	7.02 ± 0.67

were significant at 1 per cent for right hand of Arya-Thakur females which can be used for genetic investigations. The maximum observed in all communities is 47° (left hand of Arya males) and lowest value is 39° (left hand of Brahmin males). Mean values in all the three communities ranges from 41.51° to 45.50°. The variation in position of axial triradii depicted significant differences in right-left, male-female and inter-community comparisons which can be used for future investigations. A clearer picture of intra-community and inter-community variation would perhaps emerge with larger sample size. If one assumes that the sample size in the present study are larger enough, then the results would indicate that in a form not so distinct past these three communities differentiated from the same genetic stock.

CONCLUSION

This study has profiled the quantitative finger and palmar dermatoglyphic in the three communities studied in the Bhimtal Block, district Nainital, Uttaranchal. Some instances of bisexual and inter-community comparisons, which were significant at 1 per cent probability levels, do indicate that larger samples would reveal better clarity in nature of variation and hence the use in population studies of these traits. The three communities are sessile and have inhabited the area for generations. One could perhaps postulate a common origin of these groups and have resulted in the absence of larger scale significant differences in quantitative traits.

ACKNOWLEDGEMENTS

We thank Prof. R.K. Pathak, former Chairperson of the Department and currently coordinator of the UGC Special Assistance Programme, for providing necessary facilities for the present study. We also thank Ms. Himani Mankotia, who participated in the field-work which has lead to this present report. We acknowledge the cooperation of the principals, staff members and students of the schools of Bhimtal and people of Alchona and Pandeychor villages (Bhimtal Block).

BIBLIOGRAPHY

Basu, S.S., B.P. Powar and O.N. Khare. 2005. 'Palmar dermatoglyphics in pulmonary tuberculosis'. *Journal of Anatomical Society of India* 54 (2): 1-9.

Bali, R.S. and R. Chaube. 1994. *Application and methodological perspectives in dermatoglyphics*. New Delhi: Northern Book Centre.

Banik, S.D. and D.P. Mukherjee. 2006. 'Dermatoglyphics in oculocutaneous albinism'. *Anthropologist* 8 (3): 211-212.

Chakraborty, R. and K.C. Malhotra. 1981. 'Variations in asymmetry and interdigital diversity for three ridge count measures among the Dhanger caste cluster of Maharastra, India'. *Journal of Human Evolution* 10: 503- 509.

Cummins, H. and C. Midlo. 1943. *Fingerprints, palms and soles: An introduction to dermatoglyphics.* Philadelphia: The Blakiston Company.

Davee, M.A., T. Reed and C.C. Plato. 1989. 'The effect of a pattern in palmar interdigital II and a-b ridge count in black and white down syndrome cases and controls'. *American Journal of Physical Anthropology* 78(3): 439-447.

David, J.J. 1984. 'Distribution and sex variation of the a-b ridge count'. *Human Heredity* 34 (1): 14-17.

Florence, S.S., J.J. Cereghino and C.C. Plato. 1975. 'The dermatoglyphics of American Negro'. *American Journal of Physical Anthropology* 42: 183-194.

Frazer, F.C. 1994. *Development instability and fluctuating symmetry in man.* c.f. R.S. Bali and R. Chaube, 1994.

Holt, S.B. 1968. *The genetics of dermal ridges.* c.f. T.J. David, 1984.

Harich, H., E. Esteban, A. Chafik and P. Moral. 2002. 'Dermatoglyphic characterization of Berbers from Morocco: Qualitative and quantitative digital and palm data'. *Annals of Human Biology* 29 (4): 442-456.

Harvey, R.C. and D. Suter. 1983. 'Evidence for a major gene effect in the distribution of digital counts'. *Annals of Human Biology* 10 (6):565-576.

Inamdar, V.V., S.A. Val Dya, P. Kulkarni, D.B. Devarshi, S.S. Kulkarni and S.L. Tungikar. 2006. 'Dermatoglyphics in Carcinoma Cervix'. *Journal of Anatomical Society of India* 55(1): 57-59.

Igbigbi, P.S., and B.C. Masamati. 2005. 'Palmar and digital dermatoglyphic traits of Kenyan and Tanzanian subjects'. *West African Journal of Medicine* 24 (1): 26-30.

Jantz, R.L. and C.H. Hawkinson. 1979. 'Finger ridge count variability in South-Saharan Africa'. *Annals of Human Biology* 6: 41-53.

Kamali, M.S. and J. Mavalwala. 1990. 'Diversity of palmar ridge counts in Iranian population'. *American Journal of Physical Anthropology* 81 (3): 363-373.

Karmakar, B., K. Yakoveniko and E. Kobliansky. 2005. 'Complex segregation analysis of quantitative dermatoglyphics traits in five Indian populations'. *Annals of Human Biology* 32(4): 445-468.

Kohli, S. and V. Bhalla. 2000. 'Hand dermatoglyphics in deaf-mute and hare-lip and cleft palate patient'. *Man in India* 80(1, 2): 185-193.

Mahapatra, U.D. and P. Behera. 1970. 'Palmar dermatoglyphics of Mundas of Orrisa'. *Man in India* 50: 121-134.

Mate, M. 1975. 'The ridge-counts of the interdigital a-b, b-c and c-d areas in a normal sample and in cerebrally damaged patients of Thuringia, G.D.R. (East Germany)'. *American Journal of Physical Anthropology* 42(2): 233-235.

Mukherjee, D.P. 1967. *Quantitative genetics of dermal ridges with special reference to Indian population.* Ph.D Thesis, London University, London.

Murillo, P., F. Rothhammer and E. Llop. 1977. 'The Chipaya of Boliva: Dermatoglyphics and ethnic relationship'. *American Journal* of *Physical Anthropology* 46: 45-50.

Narahari, S. and J.S. Padmaja. 2006. 'Finger and palmar dermatoglyphics study among Bondos of Orrisa'. *Anthropologist* 8(4): 237-240.

Natekar, P.E., F.M. DeSouza, D.O. Motghare and A.K. Pandey. 2006. 'Digital dermal patterns in Carcinoma of breast'. *Anthropologist* 8(4): 251-254.

Newman, H.H. 1960. 'Population analysis finger and palm prints in highland and lowland Maya Indians'. *American Journal of Physical Anthropology* 18: 45-48.

Okajima, M. and K. Usukura. 1984. 'Quantitative and genetic feature of epidermal ridge minuate on the palm of twins'. *Human Heredity* 34(5):285-290.

Pateria, H.N. 1974. 'Genetics of a-b, b-c, c-d ridge counts on human palms'. *American Journal of Physical Anthropology* 40: 171-173.

Plato, C.C., J.J. Cereghino and F.S. Steinberg. 1975. 'The dermatoglyphics of American Caucasians'. *American Journal of Physical Anthropology* 42: 195-210.

Prokopec, M. and V. Sedivy. 1976. 'Australian aboriginals and the outer world: Dermatoglyphical evidence', in R.L. Kirk and A.G. Thorne (eds.), *The origins* of *Australians.* Canberra: Australian Institute of Aboriginal Studies, pp. 215-26.

Reddy, B.K.C. 2006. 'Quantitative dermatoglyphic variation among Reddi populations of southern Andhra Pradesh'. *Anthropologist* 8(4): 231-236.

Reddy, B.M., V.P. Chopra, B. Karmakar, K.C. Malhotra and H. Mueller. 2000. 'Quantitative dermatoglyphics and population structure in Northwest India'. *American Journal of Human Biology* 12(3): 315-326.

Salzano, F.M. and F. DeSaEbenevides. 1971. 'Finger print quantitative variation and asymmetry in Brazilian white and blacks'. *American Journal of Physical Anthropology* 40: 325-328.

Sunderland, E. and E. Coope. 1969. 'The Jivaro: Quantitative digital dermatoglyphics'. *Man* 4(3): 438-422.

3

FINGER DERMATOGLYPHIC VARIATION AMONG THE KUDIYAS OF COORG

K. Bhamini Raghavaiah, K. Rajasekhara Reddy
and C.E. Subramanya

INTRODUCTION

Studies on configuration of finger tips, palms and soles, that were termed as dermatoglyphics by Cummins (1926), attracted the attention of a number of personalities from different fields. Contribution of Galton has been immense in this field, covering all aspects of dermatoglyphics such as morphological classification, racial studies and inheritance which provided much incentive for development of the science of dermatoglyphics. Poll (1914) also paid much attention to the subject.

Utilization of dermatoglyphics for anthropological studies in India probably started with contribution of Biswas (1936). The subject was later enriched by Sarkar (1954). Quantitative genetics of dermatoglyphics mentioned earlier, was initiated and subsequently utilized by many well known scholars. It was found in the cited studies that the quantitative value of total ridge counts (pattern size) was inherited and determined almost entirely by one or more pairs of additive genes (Holt, 1949). The idea of utilization of quantitative value of total ridge counts was subsequently extended in the field of chromosomal aberrations. It was found that the total ridge counts showed much association with chromosomal aberrations, especially among the patients of a number of syndromic disorders such as Down, Turner and Klinfelter (Penrose, 1968).

In the present paper, an attempt has been made to study the qualitative as well as quantitative characters among the Kudiyas – the Scheduled Tribe population of Coorg district in Karnataka.

MATERIAL AND METHODS

The data for the present study was collected from Coorg district in Karnataka. Bilateral rolled finger ball prints of 200 unrelated Kudiya individuals (Males: 100 and Females: 100) were collected and analysed according to the methods and techniques given by Cummins and Midlo (1961). Appropriate statistical treatment was extended.

RESULTS AND DISCUSSION

Basic finger pattern type: The frequency distribution of basic finger pattern types (Galton) among the Kudiyas is given in Table No. 1. The study revealed high incidence of loops (50.00%) followed by whorls (42.00%) in Kudiyas in the pooled group. Further, it was also known that there was more loop variation in males (51.00%) than among the females (49.00%). The overall order of the pattern types in both sexes is L>W>A. There was no significant sex difference in Kudiyas with regard to finger ball patterns unlike in most of the populations studied earlier.

Table No. 1
Frequency Distribution of Basic Finger Pattern Types

Tribe	Sex	Number Tested	Finger Pattern Types (%)					X^2 value for bilateral difference (d.f.=2)
			Whorls	Loops			Arches	
				Ulnar	Radial	Total		
Kudiyas	M	100	42.50	45.00	6.00	51.00	6.50	
	F	100	41.50	43.30	5.70	49.00	9.50	10.48
	M+F	200	42.00	44.15	5.85	50.00	8.00	

* Statistically significant at 1% probability level

The bilateral distribution of basic finger pattern types generally showed higher incidence of loops in right hand among both males (55.40%) and females (59.60%). But the frequency of arches was higher in left hand than in right, irrespective of the sex. Radial loops were highly concentrated on the left hands in Kudiyas. The digital pattern of the population is as follows:

I : L > W > A
II : L > W > A
III : L > W > A
IV : L > W > A
V : L > W > A

SYMMETRY AND ASYMMETRY

The frequency distribution of similarity of homologous digits of right and left hands is presented in Table No. 2. The distribution of asymmetry including whorls, loops and arches was much higher than asymmetry. Males (28.00%) show a lower rate of asymmetry than the females (38.00%). The digitwise distribution of symmetry and asymmetry revealed the proportion of pattern in the order of W>L>A.

Table No. 2
Frequency Distribution of Symmetry and Asymmetry of Basic Finger Pattern Types

Tribe	Sex	Symmetry (%)				Asymmetry (%)	X^2 value for sex difference (d.f.=1)
		Whorls	Loops	Arches	Total		
Kudiyas	M	36.00	33.00	3.00	72.00	28.00	
	F	31.00	22.00	9.00	62.00	38.00	3.12
	M+F	33.50	27.50	6.00	67.00	33.00	

Monomorphic Hands

Table No. 3 depicts the frequency of monomorphic hands. Males are relatively more monomorphic (10.50%) than the females (8.50%). The frequency of monomorphic hands in both the hands is considered when all the ten fingers of an individual have the same pattern.

Table No. 3
Frequency Distribution of Monomorphic Hands

Tribe	Sex	Hand	Monomorphic Hands (%)			
			Whorls	Loops	Arches	Total
Kudiyas	M	R	5.00	6.00	0.00	11.00
		L	4.00	6.00	0.00	10.00
		R+L	4.50	6.00	0.00	10.50
	F	R	3.00	7.00	2.00	12.00
		L	4.00	1.00	0.00	5.00
		R+L	3.50	4.00	1.00	8.50

INDICES

The frequency distribution of different papillary indices based on basic finger patterns of Kudiya population like Pattern intensity index, Dankmeijer's index, Furuhata's index and Poll's index is shown in Table No. 4.

Table No. 4
Incidence of Finger Pattern Indices

Tribe	Sex	Pattern intensity index	Dankmeijer's index	Furuhata's index	Poll's index
Kudiyas	M	13.60	15.29	83.33	12.74
	F	13.20	22.89	84.69	19.38
	M+F	13.40	19.09	84.01	16.06

PATTERN INTENSITY INDEX (PII)

It is evident from the table that PII values were more or less same in Kudiyas. Males (13.60%) show slightly higher values than that of the females (13.20%).

DANKMEIJER'S INDEX

The values were higher among females (22.89%) than among the males (15.29%).

FURUHATA'S INDEX

The ratio of total whorls to total loops indicated higher values in females (84.69%) than in males (83.33%).

POLL'S INDEX

This depicted the ratio of total arches to total loops. It was higher in females (19.38%) than among males (12.74%).

DIGITWISE DISTRIBUTION OF FINGER RIDGE COUNTS (FRCs)

Digitwise distribution of mean finger ridge counts (FRCs), as shown in Table No. 5, depicted that the highest mean values of ridge counts were in I, IV and V digits in both male and female series with insignificant bilateral difference.

Table No. 5
Digitwise Finger Ridge Count (FRCs)

Digit	Sex	Hand	Ridge Count			t-value for bilaterality (d.f.=∞)
			Mean ± S.E.	S.D. ± S.E.	C.V. ± S.E.	
I	M	R	15.10 ± 0.60	6.08 ± 0.43	40.27 ± 2.84	0.31
		L	15.70 ± 0.63	6.36 ± 0.45	40.53 ± 2.86	
	F	R	11.80 ± 0.57	5.07 ± 0.35	43.01 ± 3.04	1.43
		L	12.70 ± 0.50	5.07 ± 0.35	39.92 ± 2.82	
II	M	R	11.30 ± 0.57	5.77 ± 0.40	51.07 ± 3.61	0.87
		L	11.31 ± 0.55	5.59 ± 0.35	49.51 ± 3.50	
	F	R	11.20 ± 0.52	5.24 ± 0.37	46.87 ± 3.31	2.36*
		L	10.80 ± 0.53	5.32 ± 0.37	49.30 ± 3.48	
III	M	R	11.70 ± 0.51	5.10 ± 0.36	43.67 ± 3.73	0.72
		L	11.90 ± 0.57	5.57 ± 0.40	48.55 ± 3.43	
	F	R	12.00 ± 0.48	4.89 ± 0.34	4.82 ± 2.88	0.14
		L	11.50 ± 0.55	5.54 ± 0.39	48.10 ± 3.41	
IV	M	R	14.00 ± 0.55	5.56 ± 0.39	39.76 ± 2.81	0.05
		L	13.50 ± 0.47	4.76 ± 0.33	35.33 ± 2.49	
	F	R	14.10 ± 0.49	4.91 ± 0.34	33.88 ± 2.46	1.02
		L	13.40 ± 0.50	5.04 ± 0.35	37.64 ± 2.66	
V	M	R	12.40 ± 0.52	5.21 ± 0.36	42.09 ± 2.97	1.05
		L	12.90 ± 0.47	4.75 ± 0.33	36.84 ± 2.60	
	F	R	12.50 ± 0.47	4.76 ± 0.33	38.15 ± 2.69	0.34
		L	12.40 ± 0.54	5.40 ± 0.38	43.60 ± 3.08	

* Significant at 5% probability level

DIGITWISE DISTRIBUTION OF SUMMED RIDGE COUNTS (SRCs)

Digitwise distributions of mean summed ridge counts of Kudiyas are presented in Table No. 6. The highest frequency of mean values were found in digit I and IV followed by II and V. The lowest mean values were noticed in digit III. Further, the significant bilateral difference was also observed.

TOTAL FINGER RIDGE COUNT (TFRC) AND ABSOLUTE FINGER RIDGE COUNT (AFRC)

As shown in Table No. 7, the mean values of TFRC and AFRC among the Kudiyas showed higher values among the males (129.81 and 177.30) than among the females (122.40 and 160.40).

Table No. 6
Digitwise Summed Ridge Count (SRC)

Digit	Sex	Hand	Ridge Count			t-value for bilaterality (d.f.=∞)
			Mean ± S.E.	S.D. ± S.E.	C.V. ± S.E.	
I	M	R	21.90 ± 1.11	11.19 ± 0.79	51.13 ± 3.61	
		L	19.80 ± 0.97	9.74 ± 0.68	49.21 ± 3.48	0.31**
	F	R	15.40 ± 0.92	9.26 ± 0.65	60.16 ± 4.25	
		L	17.60 ± 0.94	9.44 ± 0.66	53.67 ± 3.79	1.58
II	M	R	19.00 ± 0.94	9.43 ± 0.66	49.65 ± 3.51	
		L	15.01 ± 0.98	9.84 ± 0.69	65.22 ± 4.61	4.41***
	F	R	14.40 ± 0.86	8.69 ± 0.61	6.39 ± 4.27	
		L	16.06 ± 1.08	10.83 ± 0.76	65.28 ± 4.61	2.44*
III	M	R	15.10 ± 0.97	9.74 ± 0.68	64.54 ± 4.51	
		L	15.00 ± 0.91	9.10 ± 0.64	60.33 ± 4.26	0.10
	F	R	14.10 ± 0.81	8.13 ± 0.57	57.70 ± 4.08	
		L	15.20 ± 0.88	8.82 ± 0.62	58.08 ± 4.10	0.14
IV	M	R	20.80 ± 1.06	10.69 ± 0.75	51.41 ± 3.63	
		L	20.60 ± 1.12	11.25 ± 0.79	54.62 ± 3.86	2.36*
	F	R	18.30 ± 1.06	10.66 ± 0.75	61.63 ± 4.35	
		L	19.20 ± 0.58	8.26 ± 0.58	43.06 ± 3.04	2.24*
V	M	R	14.80 ± 0.79	7.99 ± 0.56	54.09 ± 3.82	
		L	15.20 ± 0.78	7.87 ± 0.55	51.78 ± 3.66	1.68
	F	R	14.60 ± 0.69	6.91 ± 0.48	43.37 ± 3.35	
		L	15.00 ± 0.77	7.74 ± 0.54	51.63 ± 3.65	1.34

* Significant at 5% probability level
** Significant at 1% probability level
*** Significant at 0.1% probability level

Table No. 7
Total Finger Ridge Count (TFRC) and Absolute Finger Ridge Count (AFRC) among the Medas

Tribes	Sex	TFRC			AFRC		
		Mean±S.E.	S.D.±S.E.	C.V.±S.E.	Mean±S.E.	S.D.±S.E.	C.V.±S.E.
Kudiyas	M	129.81±5.44	54.75±3.87	44.16±3.12	177.30±7.68	96.84±6.84	55.19±3.90
	F	122.40±5.15	51.25±3.62	42.87±3.03	160.40±7.34	88.74±6.27	55.32±3.91

BIBLIOGRAPHY

Biswas. 1936. Cited from Sarkar.

Cummins, H. 1926. 'Epidermal ridge configurations in development defects with particular reference to the antogenic factors which condition ridge direction'. *Am. J. Anat.* 39: 89-151.

Cummins, H. and C. Midlo. 1961. *Finger Prints, Palms and Soles: An Introduction to Dermatoglyphics.* New York: Dover Publications, Inc.

Holt, S.B. 1949. 'A quantitative survey of the finger prints of a small sample of the British Population'. *Ann.Eugen.* 14: 329-338.

Penrose. 1968. 'Medical significance of finger prints and related phenomena'. *Bri. Med. J.* 2: 321-325.

Poll. 1914. Cited from Sarkar.

Sarkar, N.C. 2000. 'Finger dermatoglyphics of the Namasudras of Nadia, West Bengal, India'. *Quantitative Genetics. J. Anthrop. Surv. Ind.*, 49: 13-39.

Sarkar, S.S. 1954. 'Finger prints in India', in *The Aboriginal Races of India, Bookland.* Calcutta, pp. 126-139.

4

FINGER DERMATOGLYPHIC VARIATION OF THE MEDAS OF COORG

K. Rajasekhara Reddy, K. Bhamini Raghavaiah
and C.E. Subramanya

INTRODUCTION

Dermatoglyphics has been widely used in population variation studies. Galton's pioneering research (1892) on the inheritance of finger print patterns stimulated many investigators. Since then standardized methods and techniques have been developed for obtaining, classifying and interpreting data. Bonneive (1924) extended the method of counting ridge to get quantitative measures of the patterns. Holt (1957 and 1968) showed the clear evidence of total finger ridge count through a small number of additive genes. Mondal and Kumbnani (1996) demonstrated that the primary (idiopathic) generalized epilepsy probably has an effect on the finger ball dermatoglyphics. Mukherjee *et al.*, 1996 studied some of the Indian populations and supported the hypothesis that a few pairs of additive alleles for high and low pattern intensities, which are compatible familiar correlations for total number of triradii on fingers. A good number of studies on finger dermatoglyphic characters are available for many populations of India (Gangadhar, 2000; Rajasekhara Reddy, 2006 and others).

The Medas, a Scheduled Tribe population of Coorg district, are mainly basket makers. This community has derived its name from the Kannada word 'bidiru', meaning bamboo. This tribal population is mainly inhabited in the hilly areas where bamboo is abundant, mainly found in the Malnadu belt. They speak the Kodagu dialect derived from Dravidian language and follow Kannada script for inter-group communications. The traditional dress of the people resembles that of the Kodavas. The family descent is patrilineal, patriarchal and patrilocal. Marriages are made through negotiation. Nuclear family is of common occurrence. Besides the household work, the Meda women also participate in political and social control affairs. Some Medas work as labourers in coffee estates and paddy fields. During festivals and marriages, they play music for the Kodavas. The drum, also called 'dudi', is the common instrument of play to the Medas. They follow the Hindu religion. They worship almost all Hindu deities. Though they rely on indigenous medicines, they also use modern medicines and follow modern health care systems. They are in favour of family planning methods. The government provides a number of facilities like employment generation schemes, mss media, transport, child welfare schemes and public distribution systems for the welfare of the people.

MATERIAL AND METHODS

The data for the present study were chosen from Coorg district in Karnataka. Bilateral finger prints of 200 individuals (Males: 100 and Females: 100) were collected and analysed according to the methods and techniques given by Cummins and Midlo (1961). Necessary statistical tools were employed.

RESULTS AND DISCUSSION

In the present study, finger pattern types according to Galton's classification with their bilateral differences by sex and also symmetry and asymmetry, monomorphic hands and the indices based on pattern frequencies were analysed. Apart from these, the quantitative characters such as finger ridge count, summed ridge count, total finger ridge count and absolute finger ridge count were also covered.

Basic Finger pattern type

Table No. 1 shows the percentage distribution of basic finger pattern types according to Galton among the Medas. Loops (53.85%) were more common in either sex of the pooled population followed by whorls (38.10%) and arches (8.05%). Further, it was also known that females (56.70%) possessed relatively more loops than males (51.00%). Thus, overall the order of the pattern types in both sexes is L>W>A. The inter-sex difference was insignificant in our study, unlike in many populations earlier studied by many investigators.

The frequency of loops in the right hand was relatively higher than in the left hand in both males and females. The frequency of arches on the left hand was more than on the right hand of either sex. However, the significant bilateral difference was noticed in female series only.

Table No. 1
Basic Finger Pattern Types among the Medas

Tribe	Sex	Hand	Finger Pattern Types (%)					X2 value for bilateral difference (d.f.=2)
			Whorls	Loops			Arches	
				Ulnar	Radial	Total		
Medas	M	R	43.20	51.40	0.40	51.80	5.00	3.56
		L	40.40	48.80	1.40	50.20	9.40	
		R+L	41.80	50.10	0.90	51.00	7.20	
	F	R	32.80	59.40	1.20	60.60	6.60	8.36*
		L	36.00	51.80	1.00	52.80	11.20	
		R+L	34.40	55.60	1.10	56.70	8.90	
	M+F	R	38.00	55.40	0.80	56.20	5.80	5.62
		L	38.20	50.30	1.20	51.50	10.30	
		R+L	38.10	52.85	1.00	53.85	8.05	

* Statistically significant at 5% probability level.

As shown in Table No. 2, the loops are highly concentrated on V and III fingers and whorls on I and IV fingers in either sex of the Medas. A general trend of the patterns of individual digits, as observed in both the sexes of the Meda tribe, is as follows:

I : L > W > A
II : L > W > A
III : L > W > A
IV : W > L > A
V : L > W > A

Symmetry and Asymmetry

Table No. 3 depicts the similarity on homologous digits of right and left hands and asymmetry in basic finger patterns. Both male and female series showed the higher proportion of symmetry including loops, whorls and arches than the symmetry with significant sex difference. In all digits except digit V, the proportion of loop symmetry is somewhat higher in either sex of the population, followed by whorl symmetry and arch symmetry (L>W>A).

Monomorphic Hands

From the frequency distribution of monomorphic hands as shown in Table No. 4, it was evident that males were more monomorphic than the females

Table No. 2
Digitwise Distribution of Basic Finger Pattern Types among the Medas

Digit	Sex	Hand	Finger Pattern Types (%)				
			Whorls	Loops			Arches
				Ulnar	Radial	Total	
I	M	R	26.50	21.50	0.00	21.50	2.00
		L	25.50	21.00	0.00	21.00	3.50
		R+L	26.00	21.25	0.00	21.25	2.75
	F	R	19.00	28.00	0.50	28.50	2.50
		L	22.50	23.00	0.00	23.00	4.50
		R+L	20.75	25.50	0.25	25.75	3.50
II	M	R	21.50	22.00	1.00	23.00	5.50
		L	18.00	19.50	3.50	23.00	9.00
		R+L	19.75	20.75	2.25	23.00	7.25
	F	R	19.00	26.50	0.50	27.00	4.00
		L	19.00	18.00	1.50	19.50	11.50
		R+L	19.00	22.25	1.00	23.25	7.75
III	M	R	14.50	33.50	0.00	33.50	2.00
		L	14.50	28.00	0.00	28.00	7.50
		R+L	14.50	30.75	0.00	30.75	4.75
	F	R	10.00	34.50	1.00	35.50	4.50
		L	10.50	31.50	1.00	32.50	7.00
		R+L	10.25	33.00	1.00	34.00	5.75
IV	M	R	30.50	18.00	0.00	18.00	1.50
		L	28.50	20.50	0.00	20.50	1.00
		R+L	29.50	19.25	0.00	19.25	1.25
	F	R	22.50	24.00	0.50	24.50	3.00
		L	26.00	21.00	0.00	21.00	3.00
		R+L	24.25	22.50	0.25	22.75	3.00
V	M	R	15.00	33.50	0.00	33.50	1.50
		L	14.50	33.00	0.00	33.00	2.50
		R+L	14.75	33.25	0.00	33.25	2.00
	F	R	11.50	35.50	0.50	36.00	2.50
		L	12.00	36.00	0.00	36.00	2.00
		R+L	11.75	37.75	0.25	36.00	2.25

Table No. 3
Digitwise Distribution of Symmetry and Asymmetry among the Medas

Tribe	Sex	Digit	Symmetry (%)				Asymmetry (%)
			Whorls	Loops	Arches	Total	
Medas	M	I	31.00	33.00	4.00	68.00	32.00
		II	24.00	29.00	11.00	64.00	36.00
		III	17.00	47.00	3.00	67.00	33.00
		IV	42.00	30.00	1.00	73.00	27.00
		V	23.00	59.00	2.00	84.00	16.00
		All	27.40	39.60	4.20	71.20	28.80
	F	I	21.00	36.00	4.00	61.00	39.00
		II	25.00	27.00	8.00	60.00	40.00
		III	14.00	59.00	8.00	81.00	19.00
		IV	36.00	32.00	3.00	71.00	29.00
		V	12.00	59.00	2.00	73.00	27.00
		All	21.60	42.60	5.00	69.20	30.80
	M+F	All	24.50	41.00	4.60	70.20	29.80

Table No. 4
Frequency Distribution of Monomorphic Hands among the Medas

Tribe	Sex	Hand	Monomorphic Hands (%)			
			Whorls	Loops	Arches	Total
Medas	M	R	7.00	8.00	0.00	15.00
		L	6.00	6.50	0.50	13.00
		R+L	6.50	7.25	0.25	14.00
	F	R	4.00	10.50	0.00	14.50
		L	5.00	7.00	1.00	13.00
		R+L	4.50	8.75	0.50	13.75

particularly on right hand among the Medas, unlike in most of the Indian studies.

Indices

The different papillary indices of the Medas such as Pattern intensity index, Dankmeijer's index, Furuhata's index and Poll's index have been presented in Table No. 5.

Table No. 5
Incidence of Finger Pattern Indices among the Medas

Tribe	Sex	Pattern Intensity Index	Dankmeijer's Index	Furuhata's Index	Poll's Index
Medas	M	13.41	17.22	81.96	14.12
	F	13.43	25.87	60.67	15.70
	M+F	13.42	21.54	71.31	14.91

Pattern Intensity Index (PII)

The pattern intensity index was considered as the number of triradii per individual, which shows deviation in its values corresponding to the incidence of higher or lower number of whorls and loops in the Medas. It was evident from the table that PII values among the males (13.4) as well as females (13.43) were more or less similar in the present study.

Dankmeijer's Index

The ratio of total arches to total whorls is known as Dankmeijer's index. This was relatively higher in females (25.87) than in males (21.54), showing more frequencies of arches in females of the Medas.

Furuhata's Index

The ratio of total whorls to total loops, considered as Furuhata's index, indicated higher value in males (81.96) than females (60.67).

Poll's Index

The ratio of total arches to total whorls was known as Poll's index. It was relatively more in females (15.70) than in males (14.12).

Digitwise Distribution of Finger Ridge Counts (FRCs)

The distribution of mean finger ridge counts along with the additive derivatives of individual fingers for each hand and sex has been furnished in Table No. 6. Right hand showed more variation than left in both male and female series among the Medas. Further, the highest and lowest values were found at I and II digits, respectively. When the combined values of right and left digits were considered, they occurred in the order of I>IV>V>III>II in both males and females.

Table No. 6
Digitwise Finger Ridge Count (FRCs) among the Medas

Digit	Sex	Hand	Ridge Count			t-value for bilaterality (d.f.=α)
			Mean ± S.E	S.D. ± S.E.	C.V. ± S.E.	
I	M	R	18.21±0.58	5.81±0.41	31.91±2.25	4.51***
		L	16.35±0.64	6.44±0.45	39.39±2.78	
	F	R	15.02±0.57	5.67±0.40	37.75±2.66	2.31**
		L	14.19±0.69	6.93±0.49	48.84±3.45	
II	M	R	13.18±0.63	6.34±0.44	48.10±3.40	2.35**
		L	12.20±0.71	7.13±0.50	58.44±4.13	
	F	R	12.93±0.62	6.16±0.43	47.64±3.36	1.90
		L	11.53±0.75	7.47±0.52	64.79±4.58	
III	M	R	14.17±0.53	5.34±0.37	37.69±2.66	1.35
		L	13.45±0.68	8.64±0.48	50.86±3.59	
	F	R	12.96±0.56	5.62±0.39	43.36±3.06	0.31
		L	12.90±0.66	6.60±0.46	51.16±3.61	
IV	M	R	15.33±0.54	5.38±038	35.09±2.48	0.86
		L	14.85±0.58	5.82±0.41	39.19±2.77	
	F	R	14.91±0.59	5.90±0.41	39.57±2.79	2.01*
		L	15.42±0.56	5.57±0.39	36.12±2.55	
V	M	R	14.47±0.46	4.60±0.32	31.79±2.24	
		L	13.91±0.49	4.90±0.34	35.23±2.49	
	F	R	13.94±0.51	5.15±0.36	36.94±2.61	3.14**
		L	14.10±0.53	5.29±0.37	37.52±2.65	

* Significant at 5% probability level
** Significant at 1% probability level
*** Significant at 0.1% probability level.

DIGITWISE DISTRIBUTION OF SUMMED RIDGE COUNTS (SRCs)

The prevalence of mean summed ridge counts along with the additive derivatives of individual fingers for each hand and sex in Medas has been portrayed in Table No. 7. The right hand was more variable than the left hand in the present study. The higher and lower mean values were found at I digit and III digit respectively in either sex of the population. When the combined values of right and left digits were considered, they occurred in the order of I>IV>V>II>III.

Table No. 7
Digitwise Summed Ridge Count (SRCs) among the Medas

Digit	Sex	Hand	Ridge Count Mean ± S.E	S.D. ± S.E.	C.V. ± S.E.	t-value for bilaterality (d.f.=α)
I	M	R	25.98±1.19	11.85±0.83	45.61±3.22	
		L	23.59±1.24	12.42±0.87	52.65±3.72	1.21
	F	R	19.12±1.02	10.15±0.71	53.09±3.75	
		L	19.00±1.10	11.04±0.78	58.11±4.10	0.14
II	M	R	18.68±1.19	11.87±0.83	63.54±4.49	
		L	16.82±1.20	12.04±0.85	71.58±5.06	0.81
	F	R	17.37±1.10	11.03±0.78	65.54±4.49	
		L	15.94±1.22	2.18±0.86	76.41±5.40	1.52
III	M	R	17.74±0.94	9.42±0.66	53.40±3.77	
		L	17.42±1.14	11.36±0.80	65.21±4.31	1.57
	F	R	15.73±0.97	9.71±0.68	61.73±4.36	
		L	15.42±1.02	10.15±0.71	65.82±4.65	1.83
IV	M	R	23.07±1.11	11.06±0.78	47.94±3.39	
		L	22.21±1.19	11.95±0.84	53.80±3.80	1.59
	F	R	20.13±1.01	10.05±0.71	49.93±3.53	
		L	21.85±1.01	10.11±0.71	46.27±3.27	2.69**
V	M	R	18.09±0.85	8.47±0.59	46.82±3.31	
		L	16.85±0.78	7.82±0.55	46.41±3.28	2.35*
	F	R	16.55±0.84	8.38±0.59	50.62±3.58	
		L	16.63±0.83	8.26±0.58	49.67±3.51	0.10

* Significant at 5% probability level
** Significant at 1% probability level

Table No. 8
Total Finger Ridge Count (TFRC) and Absolute Finger Ridge Count (AFRC) among the Medas

Tribes	Sex	TFRC Mean±S.E.	TFRC S.D.±S.E.	TFRC C.V.±S.E.	AFRC Mean±S.E.	AFRC S.D.±S.E.	AFRC C.V.±S.E.
Medas	M	146.12±4.52	45.19±3.19	30.93±2.19	200.35±8.48	84.89±6.00	42.37±2.99
	F	137.90±4.67	46.73±3.30	3.89±2.39	177.73±7.79	79.90±5.65	44.96±3.17

Total Finger Ridge Count (TFRC) and Absolute Finger Ridge Count (AFRC)

Both the mean TFRC and AFRC values were higher in the males (TFRC: 146.12 and AFRC: 200.35) than the females (TFRC: 137.90 and AFRC: 177.73).

BIBLIOGRAPHY

Bonneive, K. 1924. 'Studies on Papillary Patterns of human fingers'. *J. Genet.* 15: 1.

Galton, F. 1892. *Finger Prints*. London: Macmillan

Gangadhar, M.R. 2000. *A Genetic Study of the Adikamatakas: A Scheduled Caste Population of Mysore City, Karnataka*. Ph.D. Thesis, Mysore University, Mysore, India.

Holt, S.B. 1957. 'Genetics of Decimal Ridges: Sib Pair correlations for total finger ridge count'. *Ann. Eugen.* 20: 270-281.

——. 1968. *The Genetics of Dermal Ridge*. Springfield: C.C. Thomas.

Mondal, P.R. and H.K. Kumbnani. 1996. 'A genetic study of primary (idiopathic) generalized epilepsy with special reference to finger ball dermatoglyphics'. *Man and Life* 22 (2 and 4): 167-172.

Mukherjee, D.P., Badaruddoza and Mohammad Afzal. 1996. 'Inbreeding effects of finger prints patterns among the Qureshi Muslims and other Indian Populations'. *J. Hum. Ecol.* 7(1): 41-44.

Reddy, K. Rajasekhara. 2006. 'Finger dermatoglyphics of the Madiga – a scheduled caste of Andhra Pradesh'. *Ind. J. Phys. Anthrop. & Human Genet.* 24(1): 21-34.

5

PERMANENT TOOTH EMERGENCE STUDIES IN INDIAN POPULATIONS

V. Kaul

The knowledge about dental development has been used in different fields such as dentistry, orthodontics, physical anthropology, endocrinology, nutrition and forensic science. The growth and development of dental system has been studied along with other physiological maturity indicators, like bone age, menarche and height.

Unfortunately, study of dental development has not progressed up to the level of skeletal development, as norms for dental maturity have not been fully understood, standardized or universally accepted. The most commonly used parameter for assessment of dental maturity is dental eruption, which is easy to record, less time-consuming and less expensive. In addition, it requires rather little expertise. The term 'eruption' is often used or mostly misused for gingival or clinical appearance of the cusp of a tooth through gum.

Currently, studies on dental maturity employ histological, radiological and clinical data to assess dental development during prenatal and postnatal phase of growth. While prenatal growth is entirely based on histological investigation, postnatal growth is being investigated and correlated both by radiological and clinical data. A large number of reports are available on the chronology and timing of dental emergence. These were carried out to establish population standards for clinical, anthropological, medicolegal and other purposes. Often, these standards have been used as bases for determination of biological age.

In India, most of the studies have been conducted on emergence of permanent teeth, where invariably cross-sectional surveys have been used. Most of the studies report mean and standard deviation of time of emergence of each tooth type. Some of the studies have also reported cumulative incidence to show percentage of children, at a given age, where certain tooth or number of teeth have emerged. Currently, probit analysis has been advocated for calculation of mean age of emergence.

In India, a number of studies have been conducted since 1946 (Shourie, 1946; Nanda, 1960; Nanda and Chawla, 1966; Banerjee and Mukherjee, 1967; Sidhu and Gupta, 1973; Kaul *et al.*, 1975; Bhasin *et al.*, 1977; Nayak and Patil, 1977; Awasthi and Khare, 1978; Tiwari and Chawla, 1978; Singh, 1980; Kaul and Prakash, 1982; Jaswal, 1983; Rami Reddy, 1985; Banerjee *et al.*, 1985; Bhatnagar and Kumar, 1986; Mitra, 1986; Gaur and Singh, 1994). A perusal of literature reveals that most of the studies have been conducted in North India; only two studies each from eastern India (Bangalees of Calcutta and Tibetan of Orissa), north-east (Meitei of Manipur and Khasi of Meghalaya) and south (Madras and Gulbarga) have been reported. This clearly shows a dire need for more studies to cover different areas and populations of such a vast country. Even among the studies of North India, only two studies are reported from Himachal Pradesh (Gaddi and Kullui) and one each from Uttaranchal (Gorkha) and U.P. (Lucknow), thereby clearly showing a yawning gap in our knowledge.

A perusal of the studies reflects that methodology adopted is not uniform in all the studies. The sample selected by Shourie (1946) is a mixed population, based on school children from rice eating areas (Madras). Studies conducted in Punjab, Chandigarh, Bengal, Madras and Lucknow are based on mixed population. While Chandigarh, Punjab, Madras and Lucknow samples are from urban area, Bengalee subjects belong to the semi-rural area. Rest of the studies is community-based.

Again, different techniques have been used for the analysis of data. While most of the recent studies have used probit analysis for calculation of mean (median) age of eruption of different teeth, earlier studies (Shourie, 1946) have used mean values, whereas Banerjee *et al.* (1985), Sidhu and Gupta (1973); Rami Reddy (1985) and Tiwari and Chawla (1978) have used median values. In most of the studies, the gingival emergence of a tooth has been considered as eruption of the said tooth. It is because of this reason the two terms have often been used interchangeably. Likewise, in the present text also the term eruption stands for emergence of the given tooth. Keeping in mind the shortcomings mentioned above, the studies have been compared at different levels as detailed below:

SEQUENCE OF TOOTH EMERGENCE

Table No. 1 presents the sequence of emergence of teeth in both the jaws combined as well as separately for maxilla and mandible, for each sex, in

TABLE NO. 1

SEQUENCE OF ERUPTION OF TEETH IN DIFFERENT POPULATIONS

Population	MALE	FEMALE
LUCKNOW	$M_1\ M^1\ I_1\ I^1\ I_2\ I^2\ C_1\ P^1\ P_1\ C^1\ P^2\ P_2\ M_1\ M^2$	$M_1\ M^1\ I_1\ I^1\ I_2\ I^2\ C_1\ P^1\ P_1\ C^1\ P^2\ P_2\ M_2\ M^2$
KULU	$M_1\ P_1\ M\ P^1\ I_1\ I^1\ I_2\ I^2\ C^1\ C_1\ P^2\ M_2\ P^2\ M^2$	$M_1\ M^1\ P_1\ P^1\ I_1\ I^1\ I^2\ I_2\ C_1\ P^2\ P_2\ C^1\ M^2\ M_2$
GULBARGA	$M_1\ M^1\ I_1\ I^1\ I_2\ I^2\ P^1\ C_1\ P_1\ P^2\ C^1\ P_2\ M_2\ M^2$	$M_1\ M^1\ I^1\ I_1\ I_2\ I^2\ (C_1\ P_1)\ P^1\ M_2\ C^1\ (P_2\ P^2\ M^2)$
CHANDIGARH et al.	$M_1\ M_1\ I_1\ I^1\ I_2\ I^2\ P^1\ C_1\ P_1\ C^1\ P^2\ P_2\ M_2\ M^2$	$M^1\ M_1\ I_1\ I^1\ I_2\ I^2\ C_1\ P^1\ P_1\ C^1\ P^2\ P_2\ M_2\ M^2$
CHANDIGARH PUNJABI (KAUL)	$M_1\ M^1\ I_1\ I^1\ I_2\ I^2\ P^1\ C_1\ (P_1\ C^1)\ (P^2\ M_2)\ P_2\ M^2$	$M_1\ M^1\ I_1\ I^1\ I_2\ I^2\ C_1\ P^1\ (C^1\ P_1)\ (P^2\ P_2\ M^2)\ M_2$
PUNJABI PATIALA	$M_1\ M^1\ I_1\ I^1\ I_2\ I^2\ P_1\ P^1\ C_1\ P_2\ P^2\ C^1\ (M_2\ M^2)$	$M_1\ M^1\ I_1\ I^1\ I_2\ I^2\ P_1\ P^1\ C_1\ C^1\ P_2\ P^2\ M_2\ M^2$
HARYANA JAT	$M_1\ M^1\ I_1\ I^1\ I_2\ I^2\ (P^1\ P_1)\ C_1\ C^1\ P^2\ M_2\ P_2\ M^2$	$M_1\ M^1\ I_1\ I^1\ I_2\ I^2\ C_1\ P_1\ C^1\ P^1\ P^2\ M_2\ P_2\ M^2$
GADDI	$M_1\ M^1\ I_1\ I^1\ I_2\ I^2\ C_1\ P^1\ C^1\ P_1\ P^2\ P_2\ M^2\ M_2$	
KHASI	$M_1\ M^1\ I_1\ I^1\ I_2\ I^2\ P^1\ C_1\ P_1\ P^2\ M_2)\ P_2\ M_2\ M^2$	$M_1\ M^1\ I_1\ I^1\ I_2\ I^2\ (P^1\ C_1)\ P_1\ C^1\ P^2\ P_2\ M_2\ M^2$
MANIPURI MEITIES	$M_1\ (M^1\ I_1)\ (I^1\ I_2)\ I^2\ P^1\ P_1\ C_1\ C^1\ P^2\ (P_2\ M_2)\ M^2$	$(M_1\ M^1)\ I_1\ I^1\ I_2\ I^2\ P_1\ C_1\ P^1\ C^1\ P_2\ M_2\ P^2\ M^2$
MAGAR	$M_1\ M^1\ I_1\ I^1\ I_2\ I^2\ P^1\ P_1\ C_1\ P^2\ P_2\ C^1\ M_2\ M^2$	
GURUNG	$M_1\ M^1\ I_1\ I^1\ I_2\ I^2\ (C_1\ P^1)\ P_1\ P^2\ C^2\ P_2\ M_2\ M^2$	
BENGALIS	$(M_1\ M^1)\ I_1\ I^1\ I_2\ I^2\ M^2\ (P_1\ P^1)\ C_1\ C^1$	$(M_1\ M^1)\ I_1\ I^1\ I_2\ I^2\ C_1\ C^1\ (P_1\ P^1)\ (P_2\ P^2\ M_2\ M^2)$
PUNJABI PATIALA		$(M_1\ M^1)\ I_1\ I^2\ I_1\ I_2\ C^1\ C_1\ P^1\ P^2\ P_1\ P_2\ M_1\ M^2$
YADAV	$M_1\ M^1\ I_1\ I^1\ I_2\ I^2\ (C_1\ C^1)\ P^1\ P_1\ P_2\ P^2\ M^2\ M_2$	
GUJJAR PASSI	$M_1\ M^1\ I_1\ I^1\ I_2\ I^2\ (C_1\ C^1)\ P^1\ P_1\ M^2\ M_2\ P^2\ P_2$	
KUREEL CHAMAR	$M_1\ M^1\ I_1\ I^1\ I_2\ I^2\ (C_1\ C^1)\ P_1\ P^1\ P_2\ P^2\ (M^2\ M_2)$	

Population	MALE MAXILLA	MALE MANDIBLE	FEMALE MAXILLA	FEMALE MANDIBLE
LUCKNOW	$M^1\ I^1\ I^2\ P^1\ C\ P^2\ M^2$	$M_1\ I_1\ I_2\ C\ P_1\ P_2\ M_2$	$M^1\ I^1\ I^2\ P^1\ C\ P^2\ M^2$	$M_1\ I_1\ I_2\ C_1\ P_1\ P_2\ M_2$
KULU	$M^1\ P^1\ I^1\ I^2\ C\ P^2\ M^2$	$M_1\ P_1\ I_1\ I_2\ C\ P_2\ M_2$	$M^1\ P^1\ I^1\ I^2\ P^2\ C\ M^2$	$M_1\ P_1\ I_1\ I_2\ C\ P_2\ M_2$
GULBARGA	$M^1\ I^1\ I^2\ P^1\ P^2\ C\ M^2$	$M_1\ I_1\ I_2\ C\ P_1\ P_2\ M_2$	$M^1\ I^1\ I^2\ P^1\ C\ (P^2\ M^2)$	$M_1\ I_1\ I_2\ (C_1\ P_1)\ M_2\ P_2$
CHANDIGARH	$M^1\ I^1\ I^2\ P^1\ C\ P^2\ M^2$	$M_1\ I_1\ I_2\ C\ P_1\ P_2\ M_2$	$M^1\ I^1\ I^2\ P^1\ C\ P^2\ M^2$	$M_1\ I_1\ I_2\ C\ P_1\ P_2\ M_2$
CHANDIGARH PUNJABI	$M^1\ I^1\ I^2\ P^1\ C\ P^2\ M^2$	$M_1\ I_1\ I_2\ C\ P_1\ M_2\ P_2$	$M^1\ I^1\ I^2\ P^1\ C\ (P^2\ M^2)$	$M_1\ I_1\ I_2\ C\ P_1\ M_2\ P_2$
PUNJABI PATIALA	$M^1\ I^1\ I^2\ P^1\ P^2\ C\ M^2$	$M_1\ I_1\ I_2\ P_1\ C_1\ P_2\ M_2$	$M^1\ I^1\ I^2\ P^1\ C\ P^2\ M^2$	$M_1\ I_1\ I_2\ P_1\ C_1\ P_2\ M_2$
HARYANA JAT	$M^1\ I^1\ I^2\ P^1\ C\ P^2\ M^2$	$M_1\ I_1\ I_2\ P_1\ C_1\ M_2\ P_2$	$M^1\ I^1\ I^2\ C\ P^1\ P^2\ M^2$	$M_1\ I_1\ I_2\ C\ P_1\ M_2\ P_2$
GADDI	$M^1\ I^1\ I^2\ C\ P^1\ P^2\ M^2$	$M_1\ I_1\ I_2\ C\ P_1\ P_2\ M_2$		
KHASI	$M^1\ I^1\ I^2\ P^1\ P^2\ C\ M^2$	$M_1\ I_1\ I_2\ C\ P_1\ P_2\ M_2$	$M^1\ I^1\ I^2\ P^1\ C\ P^2\ M^2$	$M_1\ I_1\ I_2\ C\ P_1\ P_2\ M_2$
MANIPURI MECTIES	$M^1\ I^1\ I^2\ P^1\ C\ P^2\ M^2$	$M_1\ I_1\ I_2\ P_1\ C\ P_2\ M_2$	$M^1\ I^1\ I^2\ P^1\ C\ P^2\ M^2$	$M_1\ I_1\ I_2\ P_1\ C\ P_2\ M_2$
MAGAR	$M^1\ I^1\ I^2\ P^1\ P^2\ C\ M^2$	$M_1\ I_1\ I_2\ P_1\ C\ P_2\ M_2$		
GURUNG	$M^1\ I^1\ I^2\ P^1\ P^2\ C\ M^2$	$M_1\ I_1\ I_2\ C\ P_1\ P_2\ M_2$		/
BENGALIS	$M^1\ I^1\ I^2\ M^2\ P^1\ C$	$M_1\ I_1\ I_2\ P_1\ C$	$M^1\ I^1\ I^2\ C\ P^1\ (P^2\ M^2)$	$M_1\ I_1\ I_2\ C\ P_1\ (P_2\ M_2)$
PUNJABI PATIALA			$M^1\ I^1\ I^2\ C\ P^1\ P^2\ M^2$	$M_1\ I_1\ I_2\ C\ P_1\ P_2\ M_2$
YADAV	$M^1\ I^1\ I^2\ C\ P^1\ P^2\ M^2$	$M_1\ I_1\ I_2\ C\ P_1\ P_2\ M_2$		
GUJJAR PASSI	$M^1\ I^1\ I^2\ C\ M^2\ P^1\ P_2$	$M_1\ I_1\ I_2\ C\ P_1\ M_2\ P_2$		
KUREEL CHAMAR	$M^1\ I^1\ I^2\ C\ P^1\ P^2\ M^2$	$M_1\ I_1\ I_2\ C\ P_1\ P_2\ M_2$		

different populations reported. The same has been presented in graphical form. However, only those studies have been included which depict median ages calculated from probit analysis, for better understanding and clarity.

When we compare the sequence of eruption in all the studies of maxilla and mandible combined, it is evident that in the first phase of eruption of teeth, the pattern of emergence is as follows: $M_1\ M^1\ I_1\ I^1\ I_2\ I^2$. The values in subscript refer to mandible whereas values in superscript refer to maxilla. The emergence sequence reveals that mandibular teeth erupt early than their counterparts in the maxilla. Virtually the same pattern is found in both males and females in all the states, with the exception of Kullu, where it appears, mean emergence age is very early. In the second phase, when canine, premolars and second molar erupt, no clear trend is observed. However, it is apparent, among pre-molars, maxillary teeth emerge earlier than the mandibular counterparts, whereas canine and second molar invariably erupt earlier in the mandible.

COMPARISON OF TIMES OF EMERGENCE

The review of studies indicates that the first phase of teeth (i.e., first molar and incisors) start erupting at about 5 years of age and continues up to 9 years of age, whereas the second phase of teeth (i.e., canine, premolars and second molar) emerge from 10.39 years up to 13.45 years. In maxilla, the first phase of tooth eruption ranges from 5.30 years to 8.40 years in females and 5.40 to 8.85 years in males. On the other hand, in mandible it ranges from 5.10 to 7.86 years in girls and 4.09 to 7.96 years in boys. The second phase of eruption in maxilla starts from 10.00 years till 13.15 years in the males and 9.11 to 13.05 years in females. In mandible, it occurs from 9.99 to 12.66 years in the boys and 8.93 to 11.89 years in girls.

As is clear from the discription, it may not be prudent to compare individual tooth timings as different statistical tools have been applied to calculate mean values. Even when only those studies are considered where mean age by probit analysis have been calculated, no clear trends are discernible. More studies are required to arrive at meaningful conclusions based on individual tooth eruption timings.

Based on the studies available, the following inferences can be drawn:

1. No difference between left and right side of tooth eruption have been reported in most of the studies.
2. Emergence of upper and lower teeth varies according to individual tooth type. In first phase of tooth emergence generally lower first and second incisors, and lower first molar emerge early than their counterparts.

3. In the second phase, no clear trend is observed. Among premolars maxillary teeth emerge early whereas mandibular canines and second molar emerge early than their counterparts in the other jaw.
4. A lot of variation is observed in eruption pattern of teeth of the second phase.
5. Teeth erupt early in girls about 1-6 months ahead of boys, which is maximum in canines, the difference being as great as 11 months.
6. Heredity factors and ethnic origin have considerable influence over the emergence of permanent teeth. Racial differences have more significant effect on incisors and molars and less on premolars and canines.
7. Socio-economic conditions do not seem to affect emergence of teeth, but rural and urban populations do show differences.
8. Nutrition seems to have a low correlation to dental emergence, though severe under-nutrition does seem to affect skeletal as well as dental system.
9. Dental emergence is more correlated to chronological rather than skeletal age.
10. Dental emergence shows low correlation with peak height velocity or growth of face.
11. Dental emergence is an independent process and has low correlation to other maturity indicators like skeletal system, height, onset of puberty, etc.
12. From the studies available, we can safely conclude that the first phase of teeth (i.e., first molars and incisors) starts erupting at about 5 years of age and continues up to 9 years of age.
13. The second phase of teeth (i.e., canine, premolars and second molar) emerge from 10.39 years up to 13.45 years.

To conclude, more comprehensive studies are required to cover different parts of this huge country with multitude of populations to arrive at meaningful conclusions. In addition to the calculation of median ages, it is strongly felt that mean number of erupted teeth or, conversely, mean age of eruption for different number of teeth need to be reported for fruitful utilization of results for comparative and predictive analysis.

BIBLIOGRAPHY

Awasthi, S.C. and B.D. Khare. 1978. 'Sequence and age of eruption of permanent dentition among Gurkhas of Dehradun (U.P.)'. Paper presented at the 4th annual conference of Ethnographic and Folk Culture Society, U.P. held at Ahmedabad.

Banerjee, P. and S. Mukherjee. 1967. 'Eruption of deciduous teeth among Bengalee children'. *Am. J. Phys. Anthropologist* 26: 357-58.

Banerjee, P. and A.R. Banerjee. 1992. 'Eruption of permanent teeth in Bengalee school children'. *J. Hum. Ecol. (special issue)* 2: 31-36.

Bhasin, M.K., A.K. Sharma and I.P. Singh. 1977. 'Permanent dental emergence in the Kuluis of Himachal Pradesh'. *Anthropologist* 21: 9-14.

Gaur, R. and Ningthuojam Yaima Singh. 1994. 'Emergence of permanent teeth among Meiteis of Manipur, India'. *Am. J. Hum. Biol.* 6: 321-328.

Jaswal, S. 1983. 'Age and sequence of permanent teeth emergence among Khasis'. *Am. J. Phys. Anthropol.* 62: 177-186.

Kaul, S., S. Saini and B. Saxena. 1975. 'Emergence of permanent teeth in school children in Chandigarh'. *India. Arch. Oral Biol.* 20: 587-593.

Kaul, S.S., R.K. Pathak and Santosh. 1992. 'Emergence of deciduous teeth in Punjabi children'. *Z. Morph. Anthrop.* 79(1): 25-34.

Kaul, V. and S. Prakash. 1982. 'Eruption of permanent teeth in Jats of Haryana', in J. Borms *et al.* (eds.). *Human Growth and Development*. New York: Plenum Press, pp. 243-252.

——. 1989. 'Estimation of age from emergence of permanent teeth in Jats of Haryana'. *Ind. J. Phys. Anthrop & Hum. Genetics* 15(3): 109-112.

Mitra, A., C. Verma and A.K. Mitra. 1986. 'Age of eruption of permanent teeth in North Indian children'. *Ind. J. Phys. Anth. & Hum. Genet.* (2): 121-129.

Nanda, R.S. 1960. 'Eruption of Human teeth'. *Am. J. Orthodont.* 46: 363-378.

Nanda, R.S. and J.N. Chawla. 1966. 'Growth and development of Indian children. I. Development of permanent teeth'. *Am. J. Orthodont.* 52: 837-853.

Nayak, S.K. and S. Patil. 1977. 'Age of eruption of permanent teeth and its sequence among Tibetan refugees'. *J. Ind. Med. Assoc.* 69: 146-149.

Pathak, R.K., S. Jaiswal and S. Kaul. 1985. 'Association of permanent dental maturity status with stature and weight in Punjabi children of Chandigarh, India', in V. Rami Reddy (ed.). *Dental Anthropology. Application &Mmethods.* New Delhi: Inter India Publications, pp. 49-53.

Rami Reddy, V. 1985. 'Dental eruption of Indians: A Review', in V. Rami Reddy (ed.). *Dental Anthropology: Application and Methods.* New Delhi: Inter India Publications, pp. 55-73.

——. 1986. *Dimensions of Anthropology: Dentition, Pathology & Crown Morphology*, New Delhi: B.R. Publishing Corporation.

Saunders, E. 1837. *The teeth, a test of age, considered with reference to the factory children.* London: H. Remshaw.

Schour, Z. and M. Messler. 1940. 'Studies in tooth development. The growth pattern of human teeth'. Parts I & II. *J. Am. Dent. Assoc.* 27: 1778-1918.

Shourie, K.L. 1946. 'Eruption of teeth in India'. *Ind. J. Med. Res.* 34: 105-118.

Shukla, B.R.K. and B.P. Khare. 1973. 'Lucknow ke teen Samudayo ke sthai dantobhedhan (eruption) ke anukraman'. *Manav* 1(1): 29-37.

Siddhu, L.S. and P. Gupta. 1973. 'Sequence and Age of eruption of permanent teeth in the Punjabi populations of Patiala'. *East. Anthropol.* 25: 261-270.

Singh, S.P. 1980. 'Eruption of permanent teeth in Gaddi Rajput males of Dhaula Dhar range of the Himalayas'. *Z. Morph. Anthrop.* 70: 295-301.

Singhal, P. and I.J.S. Bansal. 1984. 'Relation between age at menarche, height, weight and eruptions of third molar in adult Punjabi females'. *Ind. J. Phys. Anthrop & Hum. Genet* 10(2&3): 166-171.

Tiwari, A. and A.S. Chawla. 1978. 'Study of sequence and eruption age of permanent teeth in north Indian children, Chandigarh'. *Bull. Post. Graduate Institute of Med. Edu. and Res.* Chandigarh 12: 58-68.

6

NUTRITION: ANTHROPOMETRIC PERSPECTIVE

V.S. Kulkarni and Suresh S. Alizad

A number of government and non-government organizations have made the concentrated research on the widespread malnutrition of children between 0-5 years of age. It starts as soon as breast-feeding is insufficient and acceptance of semi-solids is a changing stage. Some of the notable organizations are NCHS, UNICEF (at international level), NIN Hyderabad, NFHS, VHAI, etc. Besides, numerous small scale researches have been carried out all around. The overall conclusions and observations are that

> the infants remain relatively starved once breast milk becomes insufficient for the child's requirements around 5 to 6 months of age. There is a very close association between child malnutrition and age – and close relationships of malnutrition and the timings of receiving semisolid food. (Shanti Ghosh)

Such studies are done for the age-group 0 to 5 years children. The widespread indicators used are weight for age (Gomez Classification), height for age and weight for height (Waterloo's Classification).

According to National Family Health Survey (NFHS), knowledge and awareness regarding starting semi-solids between 5 and 6 months is scanty among families and even among health workers. NFHS results indicate that this is the beginning of malnutrition, which starts around 5-6 months of age and reaches its peak around 2 years and more or less stays at that level. With the variation in the extent, this NFHS pattern is found in all states of India.

It is a common experience of those who work among the tribals that the level of under-nutrition is higher than the level of over-nutrition: meaning thereby that because under-nutrition is more, the diseases related to over-nutrition may not be significant. Again, the environmental factors play a vital role, that is, free air, water and noise pollution. Of course, in view of the increasing communication facility through roads, media exposures and educational facilities, this statement must be taken in a restricted sense. Most of the tribal regions by and large are in a transition stage in every respect.

Anthropometry is widely recognized as one of the useful techniques to assess the growth and nutritional status of an individual or a population (Gorstein *et al.,* 1994; Jelliffe, 1966; Rao *et al.,* 1986). One of the basic reasons is that anthropometry is highly sensitive to under-nutrition (Martorell, 1984) (reproduced from Ginjinder Kaur *et al.,* 2005).

MATERIAL AND METHODS

> Physical growth is considered as one of the major outcomes of the interaction between nutrition and environment. Measurements of body weight, height, mid-upper arm circumference and triceps skin fold thickness are commonly recognized as important indices of nutritional status especially protein-energy malnutrition. Height is primarily a reflection of cumulative effect of past nutritional status whereas the other measures refer more to current or transitory nutritional status (Seoane and Latham, 1971).

The present cross-sectional study is based on a sample of 2616 individuals from 426 households of tribal population. It was collected from 67 PHCs of western Satpura region. Both male and female populations are divided into different study age-groups, viz., the population above 18 years of age was subjected to BMI whereas the population below 18 years of age was subjected to weight for age (W/A), height for age (H/A) and weight for height (W/H). The results are presented in a tabular form separately for each village. The WHO/NCHS reference and cut-off points were considered for analysis. Height and weight are the only two measurements selected. Besides, supplementary information on geographical conditions, socio-economic and health factors, food habits and food intake, etc. was collected for the assessment of nutritional status and food consumption pattern. Gomez and Waterloo classification systems were used. (WHO)

AREA OF RESEARCH

The present study was carried out by the researchers in the five selected villages from western Satpura region that addressed the question of how the problem of nutritional depletion or of malnutrition be tackled in a systematic and scientific way. The overall and the major conclusion that emerges out of the work is very much related to the improvement of life style in the villages with multifold approaches. Following is the recapitulation of the present work.

The statistical figures of 2005 with respect to the nutrient intake and the disease pattern would be the confirmation of above conclusions made couple of years ago (S.R. Chaudhary).

The nutrient intake of energy, Calcium, Niacin and Retinal, for instance is far from expected and thereby gross deficiency to the extent of more than 50% is seen. From the selected villages, however, deficiency due to the energy has not been noticed to such an extent. Could it be because of inherent sampling bias? In some of the tehsils, deficiency of fat is also noticed. The low nutrient intake and high incidence of deficiency diseases are not only disgusting but are also nerve-shattering for thinkers, planners and administrators. The downward revision in some recommended nutritional allowances and upward trend of morbidity conditions is leading to serious health and survival problems. From the analysis of mean intake of energy and associated nutrients of the present 2005 study, it seems clear that there are major deficiencies of vitamins and minerals but NOT so much of energy.

RESULTS OF ANALYSIS: BODY MASS INDEX

This is one of the best indicators of nutritional status because of its tremendous predictive capacity of nutritional, socio-economic, health and environmental features. For different villages, the extent of severe malnutrition in the total population varies. It is 48.28% for Toranmal village (27.66% for males and 20.55% for females); 25.59% for Dhavli village (13.79% both for males and female); 59.81% (29.66% for males and 30.14% females) of Surgas village; 40.53% (19.53% for males and 20.01% for females) of Satrasen (Uttam Nagar) village; and 41.89% (20.27% for males and 21.62% for females) of Gontya village. Except for village Dhavali the male children are more malnourished than the female children. This may be because of the difference in socio-economic conditions (Suresh Alizad) (Table Nos. 1 to 3). Whereas the extent of severe malnutrition for the tribals of Maharashtra has been estimated to be 14% as high prevalence (NNMB 2002). Comparing this figure with the figures of the present study, it is clear that severe malnutrition is still at the higher extent.

The frequency of LWN and the normal population together for total population of Toranmal is 14.18%, out of which male population is more in LWN (6.73%) than the corresponding female population (3.56%); whereas females are more in frequency (5.93%) in 'Normal' category than the males (2.77%); it is 43.84%, i.e., much more than Gr. III category. Males are more in frequency (12.93%) than the females (8.61%) of the same LWN category; but female frequency is more (13.28%) than the frequency of male population (13.79%) for the 'Normal' nutritional status for village Dhavli. In village Surgas, the total frequency together with LWN and Normal category is only 13.88% as against the Gr. III category. Females are more (4.30%) than the corresponding male figure (2.87%) population in LWN category. But the bi-sexual frequency is the same for Normal category (3.35%) of Nutritional Status. For village Gontya, the LWN and Normal category together for total population is 38.11% showing more incidence among females (10.0%) for LWN and even more among males (11.62%) for Normal State of Nutrition. Finally, for village Satrasen (Uttam Nagar) the frequency of LWN and Normal state of nutrition is 33.83%, i.e., less than the frequency of Gr. III condition. There is very little difference (0.59) between male and female frequencies but there is little more difference (5.32%) for Normal state of nutritional status.

Almost the same trend is observed for adult population also.

An index known as BMI is calculated for the total data as well as for adult and adolescent population (i.e., above 15 years of age). Body mass index is an index between the two body measurements, viz., the height and weight. The former is highly influenced by genetic factors and the latter by environmental factors. The first measurement is relatively constant for respective age but the second one is more fluctuating according to the changing environmental conditions. The resultant index, thus, is the predictor of genetic as well as environmental influences. The index in turn is predictive of the health conditions and the effect of related factors, socio-economic conditions and the effects of the related factors such as adequacy or inadequacy of the food and thereby the Chronic Energy Deficiency, etc. Hence, the results of BMI can be verified by the results of dietary intake-adequacy, proper-improper, etc.

HEIGHT FOR AGE, WEIGHT FOR AGE AND WEIGHT FOR HEIGHT

In the present study, a two-way comparative classification is followed, viz.;

a. Gomez classification for weight for age and
b. Waterloo's classification for height/age and weight/height.

Comparison of height for age and weight for age of the present study was made with 3rd and 50th percentile reference values for 0 to 5 years age-group children. From Toranmal village, equal number of boys and girls

Table No. 1
Sex and Village-wise BMI Classification for Total Population and Adult Population

CED Classes	Toranmal						Dhavli					
	Total Population			Adult Population			Total Population			Adult Population		
	Male	Female	Total	Male	Female	Total	Male	Female	Total	Male	Female	Total
<16	140	104	244	38	30	68	32	32	64	10	16	26
Gr.III	(27.66)	(20.55)	(48.22)	(14.96)	(11.18)	(26.87)	(13.79)	(13.79)	(27.59)	(6.75)	(10.81)	(17.57)
16.01-17.00	16	48	64	8	24	32	2	10	12	6	2	8
Gr.II	(3.16)	(9.48)	(12.65)	(3.15)	(9.46)	(12.61)	(0.86)	(4.31)	(5.17)	(4.03)	(1.35)	(5.41)
17.01-20.00	48	54	102	44	38	82	16	16	32	14	12	26
Gr.I	(9.48)	(10.60)	(20.16)	(17.32)	(14.96)	(32.28)	(6.91)	(6.91)	(13.79)	(9.46)	(8.11)	(17.56)
18.51-20.00	34	18	52	24	18	42	30	20	50	26	18	44
L.W.N.	(6.73)	(3.56)	(10.26)	(9.46)	(7.08)	(16.43)	(12.93)	(8.51)	(21.56)	(17.56)	(12.16)	(29.73)
20. 01-25.00	14	26	44	12	18	30	22	32	54	14	22	36
Normal	(2.77)	(5.93)	(8.69)	(4.72)	(7.08)	(11.08)	(9.48)	(13.79)	(23.28)	(9.46)	(14.87)	(24.32)
25. 01-30.00	0	0	0	0	0	0	10	10	20	0	8	8
Obese-I	(0)	(0)	(0)	(0)	(0)	(0)	(4.31)	(4.31)	(8.61)	0	(5.41)	(5.41)
	252	254	506	126	128	254	112	120	232	70	78	148
Total	(49.80)	(50.20)	(100.0)	(49.61)	(50.39)	(100.0)	(48.28)	(51.72)	(100.0)	(47.3)	(52.71)	(100.0)

Source: Compiled by researcher, March 2004.

Table No. 2
Sex and Village-wise BMI Classification for Total Population and Adult Population

CED Classes	Surgas						Gontya					
	Total Population			Adult Population			Total Population			Adult Population		
	Male	Female	Total	Male	Female	Total	Male	Female	Total	Male	Female	Total
<16	124	126	250	38	26	64	150	160	310	16	16	32
Gr.III	(29.66)	(30.14)	(59.81)	(19.97)	(13.54)	(33.33)	20.27	21.62	41.89	4.26	4.26	8.51
16.01-17.00	26	24	50	14	18	32	24	24	48	12	8	20
Gr.II	(6.22)	(5.75)	(11.96)	(7.29)	(9.37)	(16.67)	3.24	3.24	6.49	3.19	2.13	5.32
17.01-20.00	24	36	60	22	24	46	54	46	100	42	32	74
Gr.I	(5.75)	(8.61)	(14.35)	(11.46)	(12.50)	(32.96)	7.30	6.22	13.51	11.17	8.51	19.68
18.51-20.00	12	18	30	10	14	24	62	74	136	40	66	106
L.W.N.	(2.87)	(4.30)	(7.18)	(5.21)	(7.29)	(12.50)	8.38	10.0	18.38	10.64	17.55	28.19
20. 01-25.00	14	14	28	14	12	26	86	60	146	82	62	144
Normal	(3.35)	(3.35)	(6.70)	(7.29)	(2.87)	(13.54)	11.62	8.11	19.73	21.80	16.49	38.30
25. 01-30.00	0	0	0	0	0	0	0	0	0	0	0	0
Obese-I	(0)	(0)	(0)	(0)	(0)	(0)	(0)	(0)	(0)	(0)	(0)	(0)
	200	218	418	98	94	192	376	364	740	192	184	376
Total	(47.85)	(52.15)	(100.00)	(51.04)	(48.96)	(100.00)	50.81	49.19	100	51.07	48.93	100

Table No. 3
Sex and Village-wise BMI Classification for
Total Population and Adult Population

CED Classes	Satrasen (Uttam Nagar)					
	Total Population			Adult Population		
	Male	Female	Total	Male	Female	Total
<16	132	142	274	20	16	36
Gr.III	19.53	21.01	40.53	6.13	4.91	11.04
16.01-17.00	40	12	52	20	6	26
Gr.II	5.92	1.78	7.69	6.13	1.84	7.98
17.01-18.50	56	52	108	36	36	72
Gr.I	8.28	7.69	15.98	11.04	11.04	22.09
18.51-20.00	54	58	122	48	46	94
L.W.N.	7.99	8.58	16.57	14.72	14.11	28.83
20.01-25.00	76	40	116	60	36	96
Normal	11.24	5.92	17.16	18.40	11.04	29.45
25 01.30-00	2	10	12	0	2	2
Obese-I	0.30	1.48	1.78	0	0.61	0.61
30.01+	0	2	2	0	0	0
	0	0.30	0.30	0	0	0
	360	316	676	184	142	326
Total	53.25	46.75	100	56.44	43.56	100

Source: Compiled by researcher, March 2004.

with a percentage of about 71.87% indicate severe cases of 'stunting'; about 67% cases of Dhavali village, about 81.21% cases of Surgas village, about 48% cases of Gontya village and about 70% cases of Satrasen (Uttam Nagar) indicate 'severe stunting', i.e., the range varies between approximately 48% to 81% for all the villages. In short, the 'stunting', which is the expression of genetic endowment, is severely affected. In the same way, the 'wasting' percentages are also relatively high for all the villages. But there appears to be somewhat balanced relationship when height and weight measurements are considered together. This would indicate that conclusion based on mere height or mere weight may be misleading than the combined effect of the both. The results based only on the weight for age (Gomez's classification) would neglect the genetic expression of height. Hence, it would be recommended that for true estimation of malnourished individuals – weight

Table No. 4
Village-wise Frequency Distribution of Children (0 to 18 yrs.) – Gomez Classification

Cut-off Points/ Nutritional Status	Villages								
	Toranmal			Dhavli			Surgas		
	M	F	T	M	F	T	M	F	T
a) 90-110%	4	1	5	6	12	18	1	3	4
Normal %	2.65	0.66	3.31	8.45	16.9	25.35	0.75	2.28	3.03
b) 75- 89%	17	11	28	10	12	22	8	11	19
1st degree mild %	11.25	7.28	18.53	14.28	16.9	31.18	6.06	8.33	13.39
c) 60-74%	18	29	47	6	9	15	18	20	38
2nd degree moderate %	11.92	19.2	31.12	8.45	12.86	21.31	13.63	15.15	28.78
d) <60%	34	37	71	8	8	16	32	39	71
3rd degree severe &	22.51	24.5	47.01	121.43	11.43	22.86	24.24	29.54	53.78
Total	73	78	151	30	41	71	59	73	132
%	48.34	51.66	100	42.25	57.75	100	44.7	55.3	100

Source: Compiled by researcher, March 2004.

Table No. 5
Village-wise Frequency Distribution of Children (0 to 18 yrs.) – Gomez Classification

Cut-off Points/ Nutritional Status	Villages					
	Gontya			Satrasen (Uttam Nagar)		
	M	F	T	M	F	T
a) 90-110%	6	3	9	7	6	13
Normal %	2.88	1.44	4.32	3.5	3	6.5
b) 75- 89%	25	16	41	23	19	42
1st degree mild %	12.03	7.69	19.72	11.5	9.5	21
c) 60-74%	39	60	99	34	34	68
2nd degree moderate %	18.75	28.85	47.6	17	17	34
d) <60%	32	27	59	38	39	77
3rd degree severe %	15.38	12.98	28.36	19	19.5	38.5
Total	102	106	208	102	98	200
%	49.04	50.96	100	51	49	100

Source: Compiled by researcher, March 2004.

Table No. 6
Village-wise Frequency Distribution of Children (0 to 18 yrs.) – Waterloo's Classification

Sr. No.	Cut-off Points/ Nutritional Status	Villages								
		Toranmal			Dhavli			Surgas		
		M	F	T	M	F	T	M	F	T
	Stunting % Ht./Age									
	a) >95%	14	7	21	8	17	25	6	14	20
	Normal %	9.27	4.63	13.9	11.43	24.28	35.71	4.72	11.02	15.74
	b) 87.5-94.9%	15	21	36	2	8	10	16	21	37
	Mild %	9.93	13.91	23.84	2.85	11.43	14.28	12.6	16.53	29.13
A	c) 80.0-87%	15	24	39	9	10	19	21	19	40
	Moderate %	9.93	15.89	25.82	12.86	14.28	27.14	16.53	14.96	31.49
	d) <80%	28	27	55	10	6	16	13	17	30
	Severe %	18.54	17.88	36.42	14.28	8.57	22.85	10.24	13.38	23.62
	Total	72	79	151	29	41	70	56	71	127
	%	47.68	52.32	100	41.43	58.57	100	44.09	55.91	100
	Wasting % Wt./Ht.									
	a) >90%	34	46	80	15	17	32	13	21	34
	Normal %	26.98	36.51	63.49	30	34	64	12.04	19.44	31.48
	b) 80-89%	20	6	26	3	6	9	19	14	33
	Mild %	15.87	4.76	20.63	6	12	18	17.59	12.96	30.55
B	c) 70-79.9%	9	3	13	1	2	3	14	7	21
	Moderate %	7.14	2.38	9.52	2	4	6	12.96	6.48	19.44
	d) <70%	3	5	8	6	0	6	4	16	20
	Severe %	2.38	3.96	6.34	12	0	12	3.7	14.81	18.51
	Total	66	60	126	25	25	50	50	58	108
	%	52.38	47.62	100	50	50	100	46.3	53.7	100

Source: Compiled and analyzed by researcher, March 2004.

for height and not only weight for age – be used as an indicator. Similar comments made by Ryan *et al.* would be more thought-provoking.

> Small stature may be the reflection of both genetic make up as well as of past chronic malnutrition. There may be little functional impairment in those reduced stature but with normal weight for height, since few individuals utilized their maximum physical potential (Srikantia, 1977). People can be small and yet healthy. Safe drinking water, sanitary living

Table No. 7
Village-wise Frequency Distribution of Children (0 to 18 yrs.) –
Waterloo's Classification

Sr. No.	Cut-off Points/ Nutritional Status	Villages					
		Gontya			Satrasen (Uttam Nagar)		
		M	F	T	M	F	T
	Stunting% Ht./Age						
	a) >95%	20	10	30	19	24	43
	Normal %	9.39	4.69	14.08	9.5	12	21.5
	b) 87.5-94.9%	30	41	71	24	21	45
	Mild %	14.08	19.25	33.33	12	10.5	22.5
A	c) 80.0-87%	35	41	76	31	21	52
	Moderate %	16.44	19.25	35.69	15.5	10.5	26
	d) <80%		16	36	28	32	60
	Severe %	9.39	7.51	16.9	14	16	30
	Total	105	108	213	102	98	200
	Wasting % Wt./Ht.						
	a) >90%	56	50	106	46	36	82
	Normal %	29.63	26.46	56.09	25.99	20.34	46.33
	b) 80-89%	19	22	41	21	15	36
	Mild %	10.05	11.64	21.69	11.86	8.46	20.34
B	c) 70-79.9%	12	13	25	9	14	23
	Moderate %	6.35	6.87	13.22	5.08	7.91	12.99
	d) <70%	12	5	17	17	19	36
	Severe %	6.35	2.65	9	9.6	10.73	20.33
	Total	99	90	189	93	84	177
	%	52.38	47.62	100	52.54	47.46	100

Source: Compiled by researcher, March 2004.

conditions and personal hygiene can be more important in ensuring good health than a good weight for age, (Mahajani 1981).

NUTRITIONAL STATUS COMPARISON

The results of the present study about nutritional status have been compared with the results of the District Nutrition Profile Study (DNP), of twenty-six

Table No. 8

Comparison of Nutritional Status of Children under 3 Years of the Present Study with the Values of Various States

State*	Weight for Age Percentage Below 2SD	Height for Age Percentage Below 2SD	Weight for Height Percentage Below 2SD
Andhra Pradesh	37.7	38.6	9.1
Arunachal Pradesh	24.3	26.5	7.9
Assam	36	50.2	13.3
Bihar	54.4	53.7	21
Delhi	34.7	36.8	12.5
Goa	28.6	18.1	13.1
Gujarat	45.1	43.6	16.2
Haryana	34.6	50	5.3
Himachal Pradesh	43.6	41.3	16.9
Jammu & Kashmir	34.5	38.8	11.8
Karnataka	43.9	36.6	20
Kerala	26.9	21.9	11.1
Madhya Pradesh	55.1	51	19.8
Maharashtra	49.6	39.9	21.2
Manipur	27.5	31.3	8.2
Meghalaya	37.9	44.9	13.3
Mizoram	27.7	34.6	10.2
Nagaland	24.1	33	10.4
Orissa	54.4	44	24.3
Punjab	28.7	39.2	7.1
Rajasthan	50.6	52	11.7
Sikkim	20.6	31.7	4.8
Tamil Nadu	36.7	29.4	19.9
Uttar Pradesh	51.7	55.5	11.1
West Bengal	48.7	41.5	13.6
India	47	45.5	15.5
Sample Villages**			
Toranmal	61.25	77.58	40
Dhavli	82.05	62.5	51.28
Surgas	96.66	76.66	20
Gontya	60	55	21.05
Satrasen/Uttam Nagar	64.71	64.71	64.71

Source:

* The Indian Child: a profile 2002; table no. 14; Department of Women and Child Development, Govt. of India.

** Study of sample villages' data collected and analyzed from field-work in March 2004 and analyzed by the researcher. Data adjusted for 0 to 3 years for making comparison.

states/UTs. In Table No. 8, all these results have been presented. It is evident from this table that children under the age of three years are more undernourished, stunted and wasted than the parallel values of various states shown in the table. It indicates the need to pay more attention to improve the nutritional status of the children under the age of three years of the present study.

It would be clear that children from all these villages are lighter in weight (W/A). Frequency of lighter children varies between 60% (Gontya) and 96.66% (Surgas). They are also shorter in height (H/A) with a variation between 55% (Gontya) and 77.58% (Toranmal). Proportionately, however, the children from Surgas are less malnourished (20%) than Gontya (21.05%) followed by Toranmal, Dhavli and Satrasen (Uttam Nagar), respectively. (W/H). This may be because of the varied genetic and environmental conditions.

DETERMINANTS OF NUTRITIONAL STATUS

Another important aspect is to understand the determinants of the dietary intake and nutritional status. "In the mid 1970s few nutritive studies were available on the Indian Semi-Arid Tropical (SAT) regions of concern to ICRISAT. None of these studies had attempted to understand the determinants of dietary intake and nutritional status" (Ryan *et al.* 1987).

A number of determinants discussed positively or negatively by the concerned nutritionist is a common experience. However, the conceptual models have rarely been suggested. National agencies like NIN, ICMR, NFHS, etc. are making some efforts through their varied surveys all over. Nutritional status conceptually depends upon some direct and some indirect groups of factors. Demographic parameters, clinical and biochemical group of factors are some examples. However, quantitative results must be treated in the background of some other groups of factors that indirectly affect the nutritional status. Conceptually, these may be considered at two levels on conjuncture basic, viz., first level decides the income or purchasing capacity, occupational levels, educational levels, size of the family, health culture and the like and at the second level the agro-clinical conditions, land specification, land utilization, etc. This three-level conceptual model is suggested here on the basis of the results of the present study (Flow chart No. 1). All the three levels are to be considered jointly and not in an isolated manner.

Determinants of Nutritional Status

Genetic	Socio-economic-Cultural	Geographic
Anthropometric	Income – Cash & Kind	Agro-climatic
Clinical	Housing Conditions	Land Specification
Bio-chemical	Occupational Levels	Land Holdings
Age & Sex	Family Size, Education, etc.	Land Use
Demographic	Health Culture	Cropping Pattern
Variables	Plural Medicine	

→ Indicators – Direct <—> Indirect ←

Flow Chart No. 1: Determinants of Nutritional Status

The concept of 'protein gap' philosophy, which was a prominent strategy around 1960 to 1970, has been rejected by a number of studies afterwards (Ryan and Ashokan 1977). On the basis of the present study, even the protein energy gap philosophy may be on the verge of rejection. This has a definite base, as the result of 'nutrient intake' analysis has already been discussed and compared.

The relative importance of the agro-climatic region, features of households, socio-economic status, as assessed by anthropometric and dietary methods, would be the determining factors.

BIBLIOGRAPHY

Alizad, Suresh S. 2005. 'Nutritional Status and Deficiency Diseases in the Tribal Area of Western Satpura Region, India'. Unpublished Ph.D. Thesis, Submitted to North Maharashtra University, Jalgaon.

Census of India. 2001. *Reports.*

Chaudhari, S.R. 2001. 'Nutritional Status and Specio-Epidemiological Aspects of Deficiency Diseases in the Tribal Area of Western Satpura Region'. Minor Research Project. Report submitted to the UGC, Western Regional Office Pune.

Ghosh, Shanti. 1998. *Nutrition and Child Care: A Practical Guide.* New Delhi: Jaypee Brothers, Medical Publishers.

Gopalan, C. and K. Vijayaraghavan. 1971. 'Nutrition Atlas in India'. Hydrabad: National Institute of Nutrition.

Gorstein, J., K. Sullivan, R. Yip, M de Onis, F. Trowbride and P. Fajans. 1944. 'Issues in Assessment of Nutritional Status using Anthropometry'. *Bulletin of WHO* 72: 272-283.

Indian Council of Medical Research. 1990. 'Nutrient Requirements and Recommended Dietary Allowances for Indians'. New Delhi.

Jelliffe, D.B. 1966. 'The Assessment of the Nutritional Status of the Community'. Geneva: World Health Organization.

Kaur, Ginjinder, Harmandar Singh Kang, Pushpa Singal and S.P. Singh. 2005. 'Nutritional Status: Anthropometric Perspective of Pre-School Children'. *Anthropologist*, 7(2): 99-103.

Park and Park. 2003. *Park's Text Book of Preventive and Social Medicine.* 17th edn., Jabalpur: M/S Banarasidass Bhanot Publishers.

Ryan, J.G., P.D. Bidinger, N. Pralhad Rao and P. Pushpamma. A report (1974 and 1975); A report of the study research done for SAT (Semi-Arid Tropical) Countries – India of the Concern to International Crops Research Institute for the Semi-Arid Tropics (ICRISAT).

The Indian Child: a profile. 2002. Table No. 14. Govt. of India: Department of Women and Child Development.

World Health Organization. 'Use and Interpretation of Anthropometric Indicators of Nutritional Status'. Report of a WHO Working Group. *Bulletin of the WHO* 64(6): 929-941.

7

FERTILITY PROFILE AND FAMILY PLANNING MEASURES OF WOMEN IN STONE-CRUSHING INDUSTRY

Subir Biswas and Debopama Bagchi

INTRODUCTION

Fertility is considered as an actual level of reproductive performance in a population. However, in demography, according to Barclay (1958), "it seems most natural to view fertility as the number of births per person, or per couple, during the childbearing period of life". Again, in reality, we follow measuring fertility in relation to womenfolk of the population. It has been examined that fertility of women not only depends upon physiological functions but also some social-cultural practices. Several attempts have been made to examine association between social-cultural practices including occupation and fertility levels (Biswas and Kapoor, 2003; Chaudhury, 1996; Jennings, 1970; Visaria and Visaria, 1995).

Family planning measures, which affect fertility of women or a population, have been defined as "the ability of individuals and couples to anticipate and attain their desired number of children and the spacing and timing of their births. It is achieved through use of contraceptive methods and the treatment of involuntary infertility" (WHO, 2007). According to NFHS-2, it increased with age, education, standard of living as well as with number of living children. Nowadays, acceptance of family planning programme of India has experienced significant growth since its inception in 1951 (Santhya, 2003). National Family Health Survey (NFHS-3) shows that 64% urban and 53% rural population (overall 56.3%) in India had adopted various contraceptive methods in the year 2005-06.

The association between working status of women and fertility have been examined by various scholars. They suggested that fertility of working women is supposed to be lower than women who are confined only to household jobs (Akadli-Ergocmen, 1997; Moon, 1987; Dex *et al.*, 1998; Saxena and Aoun, 1997; Al-Riyami *et al.*, 2004; Winegarden, 1984; Gendell and Kneiter, 1970; Chaudhury, 1996). However, according to Hong (1983) women work status is not automatically related to lower fertility; and according to Singh (1988) when women's working status is not accompanied by education, it may not be associated with lower fertility. Women's work status is also associated with child loss (Nanda and Surender, 1997) as well as acceptance of family planning measures (Sundar, 1990).

In this article an attempt has been made to study the fertility performance including family planning measures of women from stone-crushing industries of Matigara Block under Darjeeling district of West Bengal, India.

MATERIALS AND METHODS

The present work was conducted during February to August 2005, on randomly selected hundred families from stone-crushing industries of Tarijot under Matigara Block of Darjeeling district, West Bengal, India. Out of 472 people of both sexes, 122 ever-married women were interviewed for the purpose. Hence, the study was supposed to analyze fertility in terms of micro-variables.

The area is concentrated by Rajbansi and different Nepali communities that migrated from Bangladesh and Nepal respectively due to political unrest; and mostly engaged in stone collecting from Balason riverbed vis-à-vis stone crushing. Among them, only Rajbansis were chosen for the study for their numerical strength and cultural homogeneity.

The analysis of fertility was carried out in period perspective, i.e., events occurring in a given period were examined. Measures of fertility rate and ratios were calculated using standard formula (Barclay, 1958; Srinivasan, 1998). Age in all respect was collected using recall method and cross-examined where needed.

RESULTS AND DISCUSSION

Various fertility measures of the population have been calculated to understand fertility trends. However, there seems to be a wide gap between actual reproductive performances (fertility) with potential level (fecundity). As it may not be possible to calculate fecundity of the population, different fertility measures have been calculated for the study.

As per Table No. 1, the child-women ratio, a simple measure of general fertility trend of the population, has been found to be 564.1. The Crude

Table No. 1
Measures of Fertility of reference Women

Fertility Measures	Value
Child-Women Ratio (CWR)	564.1
Crude Birth Rate (CBR)	38.13
General Fertility Rate (GFR)	130.43
Total Fertility Rate (TFR)	4.193
Age-Specific Fertility Rate (15-19 years)	137.93
Age-Specific Fertility Rate (20-24 years)	391.30
Age-Specific Fertility Rate (25-29 years)	142.85
Age-Specific Fertility Rate (30-34 years)	111.11
Age-Specific Fertility Rate (35-39 years)	55.55

Birth Rate (CBR), the ratio of total registered live birth per thousand populations in the year 2004, has been found to be 38.13. It has been revealed that Crude Birth Rate of the population is higher than West Bengal (20.3), India (24.8), even higher than rural areas of the respective regions (22.4 and 26.4 respectively). The General Fertility Rate (GFR), number of births per thousand women of child-bearing age (15-49 years), has been found to be 130.43, which is again higher than West Bengal (73.5), India (95.3), even higher than rural areas of these two regions (81.7 and 103.5 respectively). Age-Specific Fertility Rates or the number of births per thousand women of a given age per year, is another measures of fertility. In this study, the Age-Specific Fertility Rate has been found to be highest among women of 20-24 years age group (391.3); before that (15-19 years) it seems to be lower (137.93), and after that it declines to 142.85 for 25-29 years age group, 111.11 for 30-34 years age group and 55.55 for 35-39 years age group. Finally, it reaches zero at the age of 40 years. Whereas, for West Bengal or India it reaches its peak at 20-24 years with a value of 183.0 and 214.4 respectively and continues even till 45-49 years. Total Fertility Rate, the most important and widely used measure of fertility, may be defined as the sum of Age-Specific Fertility Rates, and calculated as 4.193, which is again much higher than West Bengal (2.3), India (3.0) even higher than their rural areas (2.5 and 3.2 respectively).

Table No. 2 (including 2A, 2B, 2C and 2D) exhibits some basic biological and social events which affect fertility levels of the population. Menarche or first menstruation is a biological event and supposed to consider as the age of puberty, which is found to be 14.26 years as mean with a wide variation of age 12 to 15 years. Age at first marriage, a social event, may be considered as functional or legal age for fertility and calculated 15.84 years as mean with a range of 13 to 21 years. The mean age at first delivery of the

Table No. 2
Biological and Social Events Related to Fertility

Events	Mean ± SD (in years)
Age at Menarche	14.26 ± 0.88
Age at First Marriage	15.84 ± 2.09
Age at First Delivery	17.13 ± 2.29
Age at Menopause	47.35 ± 3.29

Table No. 2A
Age at Menarche

Age at Menarche (in years)	No. of Women	Percentage
11-12	1	0.7
12-13	12	8.57
13-14	32	22.86
14-15	69	49.28
15-16	26	18.57
Total	140	100

Table No. 2B
Age at First Marriage

Age at Marriage (in years)	No. of Women	Percentage
11-13	5	3.88
13-15	46	35.66
15-17	42	32.56
17-19	25	19.38
19-21	9	6.97
21+	2	1.55
Total	129	100

Table No. 2C
Age at First Child-birth

Age at Mother (in years)	No. of Women	Percentage
12-14	1	0.82
14-16	47	38.52
16-18	38	31.15
18-20	19	15.57
20-22	14	11.48
22-24	1	0.82
24+	2	1.64
Total	122	100

Table No. 2D
Age at Menopause

Age at Menopause (in years)	No. of Women	Percentage
35-40	1	3.70
40-45	5	18.52
45-50	11	40.74
50-55	10	37.04
55-60	-	-
Total	27	100

Table No. 3
Family Planning Indicators

Type	Currently Using	Percentage
Oral Contraceptive Pills	20	37.74
Tubal Ligation	7	13.21
Male Condoms	15	28.30
Coitus Interrupts	3	5.66
Natural Family Planning Methods	8	15.09
Total	**53**	**100**
Contraceptives Prevalence Rate (CPR)	41.09%	

women has been observed as 17.13 years with a wide variation, starting from 14 years and may continue up to 24 years or even more. Menopause or cessation of reproductive capacity for a woman varies from 40 to 50 years or even more than that with a mean of 47.35 years.

Table No. 3 calculate the Contraceptive Prevalence Rate (CPR), the number of contraceptive user couple per hundred of married couple, as 41.09%. Among them, 15.09% of couple follow natural family planning methods, viz., abstinence and rhythm method compared to 84.91% of couple who are supposed to use modern methods. Among the users of modern birth control methods, the majority of them use oral contraceptive pills (37.74%), followed by male condoms (28.30%), tubal ligation (13.21%) and coitus interrupts (5.66%). Other forms of birth control, viz., male sterilization, intrauterine device and so on are unusual to them.

OBSERVATION

As per fertility is concerned, the fertility rates of the group remain higher as compared to the state or national level. The Total Fertility Rate (4.193) and Crude Birth Rate (38.13) of the group is higher when compared to India

(3.2 and 25.9 respectively), West Bengal (2.62 and 22.5 respectively) or even Darjeeling district (2.1 and 19.6 respectively) as per estimates of fertility from India's 2001 census. The higher fertility rates may be because of desire for more and more children, as greater number of children indicates greater number of workforce for a family. Even then, the truly childhood period is very short. Children below ten years of age have to assist their parents for stone crushing. The age-specific fertility rates have a distinctive elevation up to 20-24 age-groups, when it attains its peak (39.30) and then declines sharply, and reaches zero before the age of 40 years, compared to 50 years in case of West Bengal or India. This is mostly because of adoption of different family planning methods after that stage.

On the other hand, in respect of contraceptive prevalence rate (41.09%) it is lower than district (74.8%), state (71.2%) or even national level (56.3%) as per NFHS-3 data. Women are supposed to carry liabilities of family planning more than their men counterparts; even 13.21% tubal ligation compared to zero vasectomy.

Higher fertility of the present population may be a behavioural response to high infant mortality (which has been evident from a study on the same population by the authors; Bagchi and Biswas, 2005) of the group due to occupational hazard and others. Higher infant mortality is supposed to increase fertility level to replace the former. On the other hand, low contraceptive prevalence rate is because of their low literacy rate or educational attainment as well as less consciousness about ill effect of population growth. Therefore, the study instead of its limitation because of small size data reveals high fertility rates vis-à-vis low contraceptive prevalence rate of the women from stone-crushing industries of the region.

ACKNOWLEDGEMENTS

Authors wish to acknowledge the Centre for Women's Studies, North Bengal University for financial assistance.

BIBLIOGRAPHY

Akadli-Ergocmen, B.A. 1997. 'An Overview of Women's Status and Fertility in Turkey', in Cosio-Zavala (ed.). *Women and Families: Evolution of the Status of Women as Factor and Consequence of Changes in Family Economics.* Paris: CICRED, pp. 79-105.

Al Riyami, A., M. Afifi and R.M. Mabry. 2004. 'Women's Anatomy, Education and Employment in Oman and Their Influence on Contraceptive Use'. *Health Matters* 12 (23): 144-154.

Barclay, G.W. 1958. *Techniques of Population Analysis.* New York: Willey.

Bagchi, D. and S. Biswas. 2005. 'Women in Stone Crushing: Aspects of their Fertility Behaviour (A Micro Study from Balason Colony of Darjeeling District)'. Unpublished Report Submitted to Centre for Women's Studies, Siliguri: North Bengal University.

Biswas, R.K. and A.K. Kapoor. 2003. 'Fertility Profile of a Primitive Tribe, Madhya Pradesh'. *Anthropologist* 5(3): 161-167.

Chaudhury, R.H. 1996. 'Factors Affecting Variations in Fertility by States of India: A Preliminary Investigation'. *Asia-Pacific Population Journal* 11(2): 59-68.

Dex, S., J. Heather, S. Macran and A. McCulloch. 1998. 'Women's Employment, Traditions around Child Bearing'. *Oxford Bulletin of Economics and Statistic* 60(1): 79-98.

Gendell, M. and P.C. Kneiter. 1970. 'Fertility and Economic Activity of Women in Guatemala City, 1964'. *Demography* 7(3): 273-286.

Hong, S. 1983. 'Marital Fertility Differential by Women Working Status in Korea'. *Bull. Eugen. Soc.* 15(2): 55-60.

India, Govt. of. 2001. *Census of India*. Registrar General & Census Commissioner of India, New Delhi. From < http:// www.censusindia.gov.in > (Retrieved November 1, 2007).

Jennings, G.J. 1970. 'Cultural Factors Affecting Human Fertility'. *JASA*. 22: 52-59.

Moon, C. 1987. 'Subjective Economic Status, Sex Role Attitude, Fertility and Mother's Work'. *Ingu Pogon Nonjip* 7 (1): 177-196.

Nanda, S. and S. Surender. 1997. 'Female Work Status and its Relationship with Fertility and Child Loss in Orissa'. *The Journal of Family Welfare* 43(3): 34-37.

National Family Health Survey. 1998-99. *NFHS-2 Publications.* From <http://www.nfhsindia.org/pnfhs2.html > (Retrieved November 1, 2007).

National Family Health Survey. 2005-06. *Key Findings from NFHS-3*, From < http://www.nfhsindia.org/factsheet.html > (Retrieved May 1, 2008).

Santhya, K.G. 2003. 'Changing Family Planning Scenario in India: An Overview of Recent Evidence'. *Regional Working Papers No. 17*. New Delhi: Population Council.

Saxena, P.C. and H.Y. Aoun. 1997. 'Women's Education, Economic Activity and Fertility: Relationship Re-examined'. *Al-Abhath* 45: 25-39.

Singh, M. 1988. 'Age at Marriage and Fertility: With Special Reference to India'. *Asian Profile* 16 (1): 59-66.

Srinivasan, K. 1998. *Basic Demographic Techniques and Application*. New Delhi: Sage Publication India.

Sundar, R. 1990. 'The Status of Women and Family Planning Acceptance: Some Field Results'. *The Journal of Family Welfare* 36: 60-68.

Visaria, L. and P. Visaria. 1995. 'India's Population Transition'. *Population Bulletin* 50(3): 1-51.

Winegarden, C.R. 1984. 'Women's Fertility, Market Work and Marital Status: A Test of the New Household Economic with International Data'. *Economica* 51: 447-456.

World Health Organization. 2007. *Working definition used by the WHO Department of Reproductive Health and Research*. From < http://www.who.int/topics/amily_planning/en/ > (Retrieved November 1, 2007).

8

BODY SIZE AND MENARCHEAL STATUS OF ADOLESCENT GIRLS OF NAHAN, HIMACHAL PRADESH

Taranpreet Kaur and R.K. Pathak

INTRODUCTION

Puberty marks a period of swift growth that is manifested in body size, shape and composition. This phenomenon of adolescent growth spurt occurs in all children though it varies in intensity and duration from one child to another. The adolescent spurt in skeletal and muscular dimensions is closely related to rapid development of the reproductive system which takes place at this time. There is a close relationship between age at menarche and age at PHV; indeed, it is perhaps the closest relationship between any two events occurring in different systems during puberty. The correlation is typically between 0.8 and 0.9. Menarche occurs on an average about 1.3 years after PHV, with a range of 0-2.5 years. Only 1% of girls, if that, have menarche before PHV (Marshall And Tanner, 1986). Further, all girls start to menstruate when the height velocity is falling, i.e., during downward part of the height velocity curve. Indeed, on an average menarche occurs at the time of maximum deceleration of height growth, the moment the velocity is dropping fastest (Tanner, 1978). Simmons and Greulich (1943) showed that earlier the menarche, the greater the peak velocity of height spurt. It seems that in earlier maturers, the whole process goes more quickly and more intensely so that a greater total result is achieved despite the smaller amount of time taken (Tanner, 1962). Ellison (1982), too, confirmed that height velocity was a far better predictor than weight, suggesting that the timing of menarche is determined by the rate of maturation and not body size. Twin studies also show that genetic control of sexual maturation is independent of corresponding genetic system determining the body size (Loesch *et al.*, 1995)

Contradicting above, Sharma *et al.* (1988) reported that girls with early menarche have significantly smaller skeletal dimensions and more subcutaneous fat than those with average and late menarche, indicating a biological determinant of the observed association between menarcheal age and fatness and body size. Similarly, Frisch and Revelle (1970) related the age at menarche to the attainment of critical body weight. Karen *et al.* (2002) too suggested the important role of increased body fatness on menarche.

Age at menarche is positively related to both body weight and height (Helm *et al.*, 1995; Bagga and Kulkarni, 2000) and it is found that pubescent girls are heavier and taller than non-pubescent girls.

With the above background, the present study is conducted on a cross-sectional sample of 286 school-going girls aged 11-16 years of Nahan area in Sirmaur District of Himachal Pradesh to

(i) report the average for height-weight and menarche and
(ii) understand the relationship of menarche with height and weight.

MATERIAL AND METHODS

Nahan is situated over an isolated ridge in the Siwaliks, at an altitude of about 932 metres. Nahan, founded by Raja Karam Prakash in AD 1621, is presently the district headquarters of Sirmaur district. It has several famous temples and tanks which add considerably to its attractions, with an ancient palace. The climate is pleasant here throughout the year. Nahan town has a total population of 26,053 persons, of which 46.33% are females (Census of India, 2000).

The main dialect of the area is 'Pahari Sirmauri' followed by Hindi. The Hindus predominate the region with traces of Sikhs and Muslims along with Jain and Buddhist communities. The Rajputs form the dominant (in number) caste group followed by Kolis and Brahmins. Other caste groups found in the area are Baniyas, Domes and Chimbas.

Before actual data collection, an attempt was made to establish rapport with local people and to get familiarized with the area and with the way of life of people. Elders, school teachers and some of the subjects were explained the purpose and nature of information required for the present study. The data were collected from the following schools located in the area: Govt. Middle School, Dadahu; Govt. High School, Chandani; Govt. Senior Secondary School, Sangrah; and Govt. High School, Deed Panar.

Each subject was measured for height and weight and information on menarcheal status was also sought. Other general information of each subject, including name, date of birth, caste, education and occupation of parents, number of family members, total income, etc. were also recorded. The date of birth of each subject was also checked from school records. In cases where the two dates of birth did not match, such subjects were excluded from the study. Only those subjects, who were apparently physically and mentally normal and did not suffer from any deformity, were included in the sample. The age of the subjects were converted into decimal age using the decimal age calendar given by Tanner *et al.* (1966).

Arithmetic mean, standard deviation and Student's t-test have been employed to interpret the data in numerical terms. The results thus obtained are presented in tabular as well as graphical forms. Information about age at menarche was obtained by 'status-quo' method. Further, statistical tables by Fisher and Yates (1957) were put to use for probit analysis of menarcheal data of girls of present study. The percentage of girls who had experienced menarche was calculated and was converted to probit values. Then median age was calculated using the graphical method with age on X-axis and corresponding on Y-axis.

RESULTS

By looking at the tabular and graphical representation of the data analytically, it can be said that there is a linear relationship between height and age, weight and age. The mean values of the data on height show that 11-year-old girls on an average are 135.39 cm tall. From 11 to 16 years of age, the girls of present study show a total gain of 21.26 cm in their height. Out of this total gain, maximum of 27% (5.87cm) occurs in the first year (i.e., 11 to 12 years), followed by 24% from 13 to 14 years of age. Considering the total sample of the present study, the shortest and the tallest girls measure 121 cm and 164.8 cm, respectively.

Weight shows a steady gain in mean value from 11 to 16 years of age with a total gain of 13.98 kg. Out of this, maximum gain of 5.45 kg (38.98%) is observed from 13-14 years of age followed by 4.62 kg (33.05%) from 11-12 years of age. Between 12-13 years' girls of the 'Renuka' is the most important place of religious interest in Sirmaour district. It is nearly 40 km away from Nahan. This place has a famous lake, which vaguely resembles a human figure. It is believed that Lord Parshurama's mother consecrated into water out of which this lake was formed. This famous lake is visited by thousands of pilgrims every year on *Karlika ekadashi,* when the brass idol of Parshurama is brought from village Jamu, his permanent abode, in a silver *palki* accompanied by musical instruments with great pomp and show. The deity remains in the temple for three days, i.e., *Sudi dashmi* to *Dwadashi.*

Clothing as such bears no specific difference from surrounding states. Females were spotted wearing *salwar-kameez*; males were wearing *kurta-pyjama* or pant-shirt. Ladies in rural area tie a long scarf over their head.

Mostly people are vegetarian and eat seasonal vegetables like potato (most popular one), brinjal, cabbage, cauliflower, etc. along with pulses like *moong, channa, rajmah* and *urad*; wheat, maize and rice form the staple food of the population. 'Khir Patanda' is a special dish made on the *Purnima* and special occasions. *Makki* (roasted corn) and guava are the favourite snacks eaten between the meals. Milk also forms part of the diet but only in those families who are engaged in dairy farming or of high socio-economic status.

Agriculture constitutes the main occupation of the population followed by government jobs (mostly class IV employees). Some are teachers and a smaller number of the population draws income from business (small hotels or shops, etc.) or private jobs in surrounding towns while some others are engaged in miscellaneous jobs like LIC agents and other skilled jobs (mason, carpenter, etc). Females are mostly engaged in managing house chores or assisting in farming with a fewer ones going for professions like teaching; others are helpers in Anganwadi centres or working as house-maids.

Polygamy is also observed in the region. The increased work load, assistance in agriculture or longing for son are the reasons given for polygamy. The families not only vary in size but in structure also, with both joint and nuclear kind of families in existence. The number of children varies from 1 to 11 with average 4 children per family. Although health care facilities, both government and private, are available in the region but the belief in the traditional magical and religious practices is more popular in the population, as every second subject was seen with a sacred thread tied around the neck for cure of some disease. Similarly, educational facilities

Table No. 1
Age-wise Statistics for Mean Height and Weight

Age Group (Years)	**N**	**Height (cm) Mean±SD**	**Weight (kg) Mean±SD**
11	48	135.39±5.91	27.25±4.69
12	55	141.26±6.93	31.87±6.03
13	52	143.74±7.59	32.61±6.13
14	39	148.84±5.92	38.06±4.32
15	43	152.25±7.12	39.07±5.76
16	49	156.65±5.35	41.23±4.76

at primary, middle and secondary level are available. However, it is observed that many children, especially girls, drop out of school after middle standard.

The present study shows minimum gain (0.74 kg) in body weight. Individual body weight among subjects of the entire sample of 286 girls varies between 19 and 60 kg.

The median age (±SD) at menarche for adolescent girls of the present sample is found to be 13.90 ± 0.85 years.

According to Tanner (1978), on average menarche occurs at the time of maximum deceleration of height-growth, the moment the velocity is dropping fastest. To test this hypothesis, the data on height-growth is plotted on acceleration curve, and then the median age of menarche is marked on the curve. It is evident from Figure 3 that menarche coincides with the maximum deceleration of height spurt.

To understand the relationship of menarche to height and weight, the sample in each age-group has been divided according to the occurrence and non-occurrence of menarche, and then mean ± SD for height and weight has been calculated. Further, Student's t-test has been applied to the resultant data in order to evaluate the statistical significance of differences between the two groups, and the same in presented in Table No. 2.

It is clear from the table that in each age-group, girls in whom menarche had occurred possess greater mean height as compared to their counterparts in whom menarche had not occurred. However, statistically significant

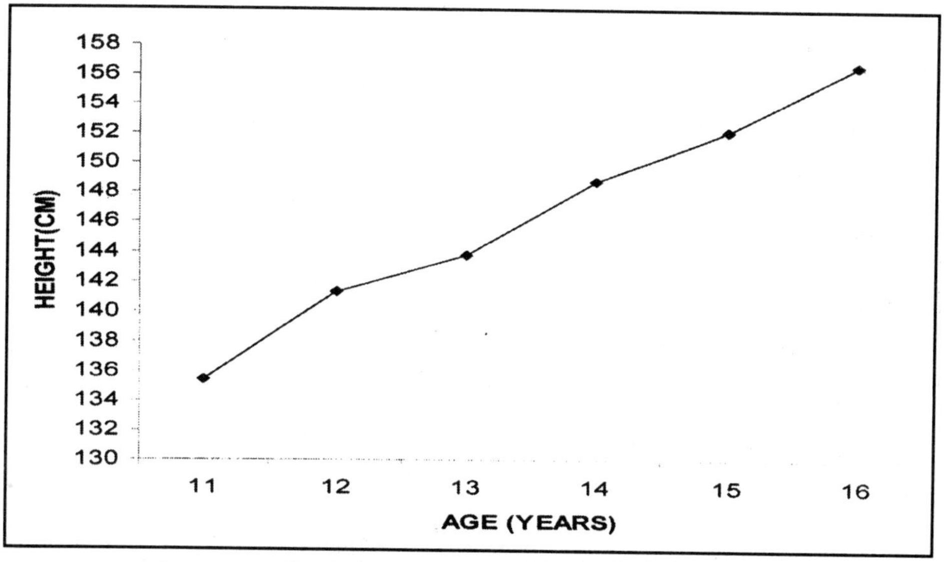

Fig. 1. Distance Curve for Height.

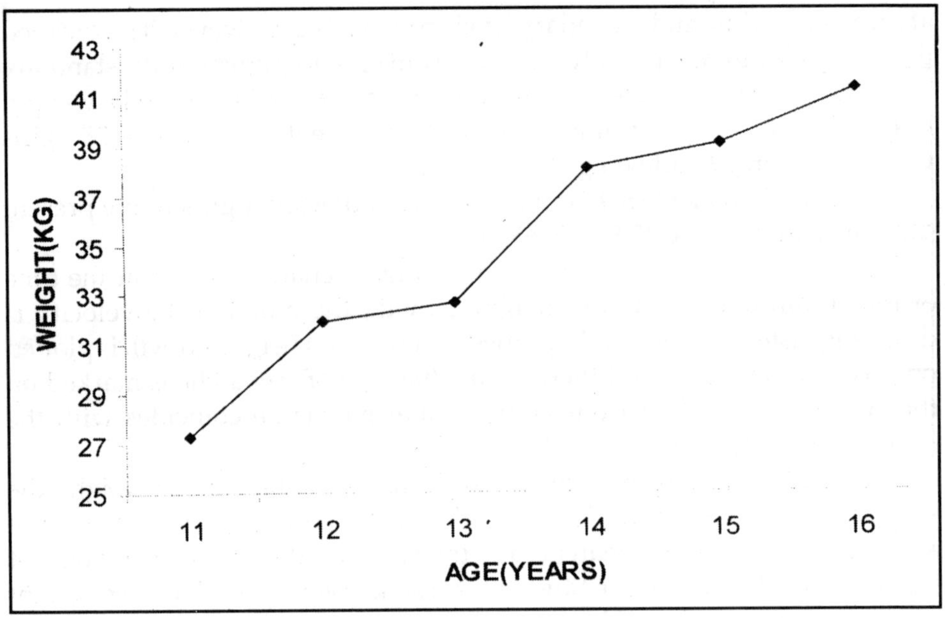

Fig. 2. Distance Curve for Weight.

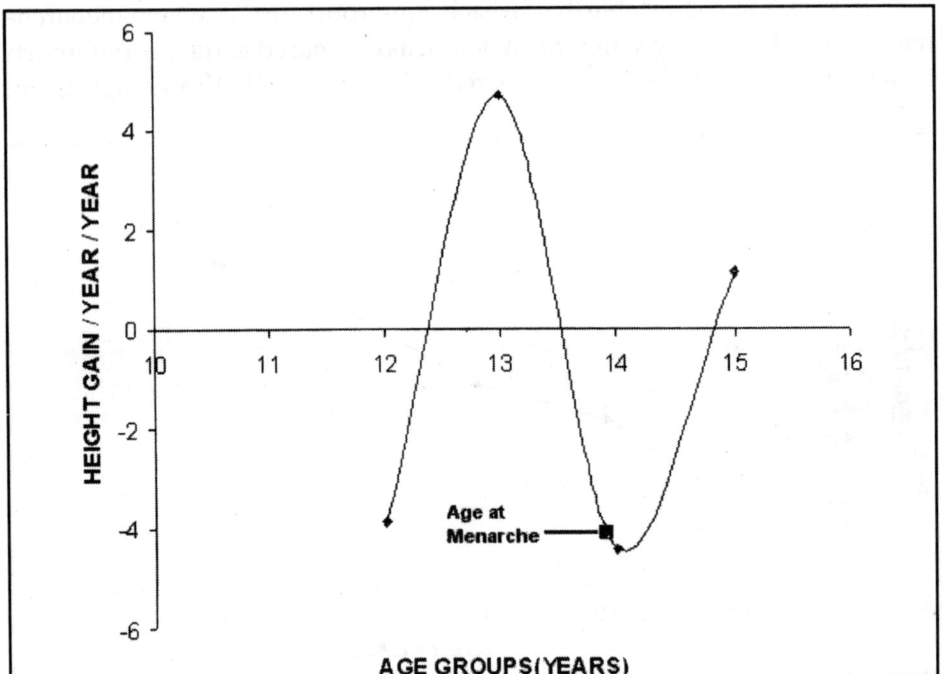

Fig. 3. Acceleration Curve for Height in Relation to Menarche.

Table No. 2
Relationship of Menarche to Height and Weight

Age	Height (cm.)				t-values	Weight (kg.)				t-values
	Menarche occurred		Menarche not occurred			Menarche occurred		Menarche not occurred		
	N	Values	N	Values		N	Values	N	Values	
11	1	145 (-)	47	135.20 (5.81)	-	1	43.00 (-)	47	26.90 (4.13)	-
12	10	141.38 (6.82)	45	140.73 (7.33)	0.256	10	33.80 (6.55)	45	31.20 (5.86)	1.106
13	9	153.59 (3.28)	43	142.68 (6.16)	4.942**	9	42.11 (4.51)	43	30.62 (4.33)	6.642**
14	21	149.32 (5.90)	18	148.28 (5.89)	0.533	21	39.71 (3.72)	18	36.14 (4.08)	2.900*
15	32	153.72 (6.01)	11	147.96 (8.29)	2.035*	32	40.72 (5.03)	11	34.27 (5.02)	3.449**
16	43	152.99 (5.39)	6	150.15 (4.66)	1.267	43	41.74 (4.75)	8	37.50 (3.08)	2.718*

** Significant at 1% level
* Significant at 5% level

differences in mean height are observed only among girls belonging to the age-group of 13 and 15 years.

A perusal of mean values for weight, presented in the table, clearly indicates the same trend as that observed for the mean height. That is, girls in whom menarche had occurred are heavier than the ones in whom menarche had not occurred. But unlike height, mean weight shows statistically significant differences at all age-groups considered, except at 12 years.

The findings of the present study clearly indicate the relationship between menarche and body size (height and weight) and are in confirmation with those given by Frisch and Revelle (1970), Tanner (1978) and Karen *et al.* (2002).

BIBLIOGRAPHY

Bagga, A. and S. Kulkarni. 2000. 'Age at menarche and secular trend in Maharashtrian girls'. *Acta Biologica Szegediensis* 44: 53-57.

Ellison, P.T. 1982. 'Skeletal growth, fatness and menarcheal age: A comparison of two hypotheses'. *Human Biology* 54: 269-281.

Fisher, R.A. and F. Yates. 1970. *Statistical tables: For biological, agriculture and medical research*. London: Oliver and Boyd.

Frisch, E.R. and Roger Revelle. 1970. 'Height and weight at menarche and a hypothesis of critical body weights and adolescent events'. *Science* 169: 397-399.

Harrison, G.A., J.M. Weiner, J.M. Tanner, N.A. Barnicot and V. Reynolds. 1977. *Human biology: An introduction to human evolution, variation, growth and ecology*. Oxford: Oxford University Press.

Helm, P., K. Munster and L. Schmidt. 1995. 'Recalled menarche in relation to infertility and adult weight and height'. *Acta Obstet. Gynecol. Scand.* 74: 718-722.

Karen Lin Su, M.G. Vogiatzi and M.I. New. 2002. 'Body mass index and age at menarche in adolescent clinic population'. *Clinical Pediatrics* 41 (7): 501-507.

Loesch, D.Z., R. Huggins, E. Rogucka, N.H. Hoang and J.L. Hopper. 1995. 'Genetic correlate of menarcheal age: A multivariate twin study'. *Annals of Human Biology* 22: 479-420.

Marshall, William A. and James M. Tanner. 1986. 'Puberty', in Frank Falkner and J.M. Tanner (eds.). *Human Growth: A comprehensive treatise*, 2nd edn., Vol. 2. New York: Plenum Press, pp. 171-205.

Sharma, K., I. Talwar and N. Sharma. 1988. 'Age at menarche in relation to adult body size and physique'. *Annals of Human Biology* 15(6): 431-434.

Simmons, K. and W.W. Greulich. 1943. 'Menarcheal age and height, weight and skeletal age of girls age 7 to 17 years'. *J. Pediatrics* 22: 518-548.

Tanner, J.M. 1962. *Growth at adolescence.* USA: Oxford Scientific Publications.

——. 1978. *Foetus into man: Physical growth from conception to maturity.* Cambridge: Harvard University Press.

Tanner, J.M. and R.H. Whitehouse. 1982. *Atlas of children's growth.* New York: Academic Press.

9

NUTRITIONAL STATUS AND ORAL HYGIENE OF RAJPUT BOYS OF THEOG AREA, HIMACHAL PRADESH, IN RELATION TO PERMANENT TOOTH EMERGENCE

Rajan Gaur, Teena Singal and Suresh Kumar

INTRODUCTION

Nutritional status is one of the most important components of health of a community. Poor nutritional status has been found repeatedly to be closely related to increasing risk of death among children in the developing countries of Africa, Asia, and Latin America (Mclaren *et al.*, 1969; Chen *et al.*, 1980; Briend *et al.*, 1986; Fauveau *et al.*, 1990; Vella *et al.*, 1994; Garenne *et al.*, 2000). Malnutrition is the main risk factor for mortality in the world, contributing to nearly 12 per cent of all deaths and 16 per cent of all disability-adjusted life years lost globally (Murray and Lopez, 1997) and one-third of all disability-adjusted life years lost in low-income countries (Mason *et al.*, 2003). Among the people living below the poverty line in the developing world, malnutrition is one of the largest causes of morbidity and premature deaths (May and Montagu, 2008). A consistent relationship between child mortality and low weight-for-age has been reported (Pelletier *et al.*, 1994, 1995). It has been estimated that, in the developing countries, 56% of deaths in children below the age of five years were attributable to malnutrition, and that 83% of this total was due to mild and moderate forms (Pelletier *et al.*, 1994, 1995). Under-nutrition, directly affects many aspects of the development of children. In particular, it retards their physical and cognitive growth and increases susceptibility to infection and disease, further increasing the probability of being malnourished (Gragnolati *et al.*, 2005). Moreover, undernourished children fail to grow up to their full genetic potential. Protein-energy malnutrition in early childhood has also been found to be associated with poor cognitive and motor development (Grantham-McGregor and Ani, 2001), which in turn undermine educational attainment. This can ultimately have an impact on productivity at work and at home, with adverse implications for the economic growth of a nation.

India is now one of the fastest growing economies in the world with gross domestic product (GDP) growth over 8 per cent. There has been a steady but slow decline in poverty (Ramachandran, 2007). But, paradoxically, despite rapid economic growth in the last decade, India is home to 40 per cent of the world's malnourished children and 35 per cent of the developing world's low-birth weight infants (IFPRI, 2008). It is a matter of concern that anthropometric indicators of nutrition in India, for both adults and children, are still among the worst in the world. The prevalence of underweight among children in India is amongst the highest in the world, and nearly two times that of Sub-Saharan Africa (Gragnolati *et al.*, 2005). According to recent estimates of the National Family Health Survey (NFHS-3), in 2005-06, 46 per cent of children under three years of age were underweight, 38 per cent were stunted, 20 per cent severely malnourished and 10 per cent were wasted (IIPS, 2007). There is also large inter-state variation in the patterns and trends in underweight (Gragnolati *et al.*, 2005). It is therefore important to have an estimate of the type and extent of malnutrition prevailing in a region for an intervention programme to have any success. Research investigations designed to evaluate the nutritional status of a community serve a valuable need by providing the necessary information for policy planners.

Besides being the single most important cause of child mortality, malnutrition severely retards a child's cognitive, physical and emotional growth and has a cascading effect on his/her productivity in adult years. In addition, it is also believed to affect the development, emergence and carious condition of dentition (Mellanby, 1928; Schour *et al.*, 1941; Shaw, 1963; Alvarez and Navia, 1989; Sweeny *et al.*, 1969; Alvarez, 1995). Some studies have investigated the relationship between tooth emergence and weight and stature (Bailit and Sung, 1968; Seow, 1996; Billewicz and McGregor, 1975; Kaul *et al.*, 1975; Visweswara Rao, 1985; Pathak *et al.*, 1985; Banerjee *et al.*, 1992). However, only a few studies have investigated the relationship between under-nutrition and tooth emergence (Niswander, 1963; Triratana *et al.*, 1990; Ulizaszek,1996; Visweswara Rao, 1992) and the issue is yet to be adequately resolved.

The present investigation aims to investigate the nutritional status of 6 to 13 year old Rajput boys of Theog area of Shimla district in Himachal Pradesh State. The study also investigates the effect of under-nutrition on permanent tooth emergence among the Rajput boys of this area.

MATERIALS AND METHODS

The present investigation is based on a cross-sectional sample of 202 Rajput boys ranging in age from 6 to 13 years. The data were collected from the various schools located in the Theog area (Fig. 1) of Shimla district of Himachal Pradesh state during September 2004. The following eight schools were visited for data collection:

1. Rahighat Primary School, Rahighat
2. Blue Bell School, Theog
3. Government Lala Roshan Lal Primary School, Theog
4. Government Senior Secondary School, Theog
5. Deori Ghat Primary School, Deori Ghat
6. Central Primary School, Matiana
7. Saraswati Vidya Mandir, Theog
8. Jawahar Navodya Public School, Theog

Only normal healthy subjects were included in the sample. Dates of birth of the subjects were obtained from the school record and, in case of doubt, verified from the subjects, their sibs and teachers. The age of each subject was converted to decimal age using the 'Decimal Age Calendar' of Tanner and Whitehouse (1966). The subjects were divided into eight age-groups of one year each using age mid-points. The number of children was 23, 21, 24, 23, 23, 31, 32 and 25 in age groups 6, 7, 8, 9, 10, 11, 12 and 13, respectively.

The nutritional status of each subject was assessed with the help of stature and weight, which are the two basic measurements for the assessment of nutritional status (WHO, 1986). Stature and weight of each individual was recorded using standard instruments and techniques following Weiner and Lourie (1981). As per the recommendations of the World Health Organization (WHO, 1983), the following derived indices were calculated for each individual for the assessment of nutritional status:

1. Weight-for-Age
2. Stature-for-Age
3. Weight-for-Stature

The reference data of the National Centre for Health Statistics, USA (Hamill *et al.*, 1979) was used to assess the nutritional status, following the recommendations of the World Health Organization (WHO, 1981). For each individual, the nutritional status was calculated as Z-scores or SD scores, following Waterlow *et al.* (1977). The cut-off point for malnourished children was taken as – 2 SD scores below the reference median, as recommended by the World Health Organization (WHO, 1983). Children with SD scores

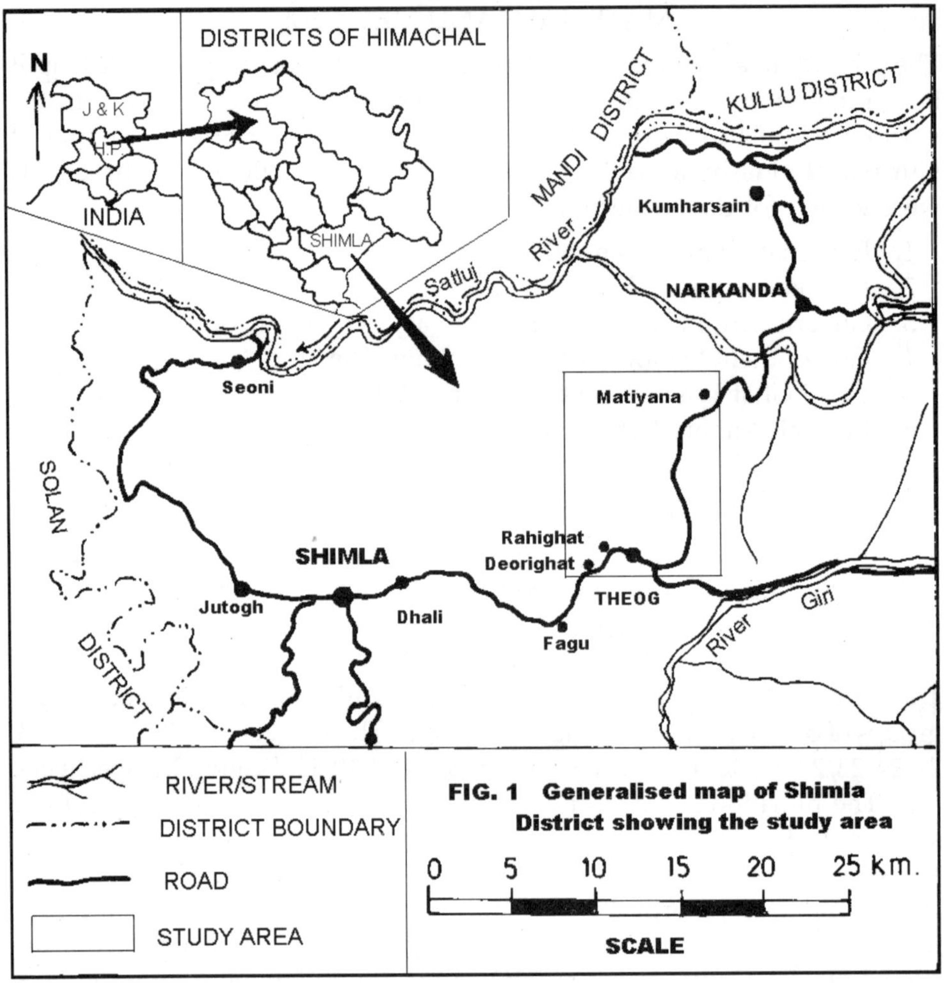

Fig. 1.

of –1 or over were included under the category of normal children. The relationship of total number of emerged teeth with stature and weight was calculated using partial correlation.

To obtain information on the number of emerged teeth, each subject was examined in day light for the emergence of permanent dentition. A tooth was recorded as emerged if any part of the tooth crown had pierced the gingiva and was visible in the oral cavity. Whenever necessary, dental probe, dental mirror and torch were also used. Apart from the dental emergence data, a record of carious teeth of each subject was also kept. In addition, information about some aspects of oral hygiene, such as type of dentifrice used and regularity of its use, was also recorded on a proforma specially designed for this purpose. Median ages of emergence of various permanent teeth were calculated using probit analysis (Mayhall *et al.*, 1978).

Theog is an important town of the area and is the headquarter of a tehsil with the same name. It is located on the National Highway-22 at a distance of 22 kilometres from Shimla town and about 8 kilometres from Narkanda town. The area is mountainous and the altitude varies from 3000 feet to 8000 feet above mean sea level. The temperature varies from -2° C in winters to about 29° C in summers. Theog tahsil has an area of 33 square kilometres and a total population of 77,967 persons (39,918 males and 38,049 females) with a sex ratio of 953 (Census of India, 2001).The upper reaches of the area experience moderate to heavy snowfall during winter. The main caste groups of the area are Brahmins and Rajputs. The staple food of the people of the area includes maize, rice and wheat.

RESULTS AND DISCUSSION

General Aspects

Family type and family size: A majority (53.8%) of the subjects in the present sample were observed to be from a nuclear family set-up. Joint family (39.2%) was the next most common family type and the percentage of subjects living in an extended family set-up was comparatively low (7.0%). The sample was divided into five categories according to the number of members in a family, i.e., those having up to 4, 5 to 8, 9 to 12, 13 to 16, and more than 16 members. A large majority of the families in the present sample had either up to 4 (33.4%) or 5 to 8 (48.7%) family members. The percentage of large families, i.e., with family size of 13 or over, was very low (4.5%).

Education of parents: Table No. 1 shows the distribution of parents of the Rajput boys of Theog area in different educational categories. It is clear

Table No. 1
Percentage Distribution of the Parents of the Rajput Boys of Theog Area in Different Levels of Education

Education Level	Father (%)	Mother (%)
Illiterate	12.9	17.3
Up to Primary	23.3	26.7
Primary to Below Matric	9.4	21.8
Matric up to Graduation	43.6	32.2
Graduation & Above	10.8	2.0

from the table that the extent of illiteracy was more among the mothers (17.3%) than the fathers (12.9%). The percentage of fathers (54.4%) who had education beyond matriculation was distinctly more than that of the mothers (34.2%). The percentage of graduate parents showed a gender bias in favour of the fathers (2% mothers and 10.8% fathers were graduates).

Oral Hygiene

The results indicate that the Rajput boys in the present sample had an average level of oral hygiene. The incidence of caries was not very high. Nevertheless, 13.36% of the Rajput boys had one or more carious permanent tooth. A majority (73.26%) cleaned their teeth regularly while about one-fourth (26.74%) were irregular in cleaning their teeth. Tooth paste (96.5%) was observed to be the most commonly used dentifrice. Some Rajput boys (2.47%) used 'Datun' and a negligible percentage (0.99%) used tooth powder.

Nutritional Status

For the assessment of nutritional status, anthropometric method was used and derived indices of stature and weight were calculated. It is one of the basic techniques for the comprehensive evaluation of nutritional status (Johnston, 1981). The stature and weight of the Rajput boys of the present sample were expressed as S.D. scores of stature-for-age, weight-for-age and weight-for-stature using the reference standards of National Centre of Health Statistics, USA (Hamill *et al.*, 1979). Table No. 2 shows the percentage of Theog boys classified in different nutritional categories expressed as S.D. scores. Following the recommendations of the World Health Organization (WHO, 1983), the boys with S.D. scores of -2 or under were classified as malnourished. It is evident from the table that 34.7% of the boys were under-

Table No. 2
Incidence (%) of Malnutrition according to Weight-for-age (Wt/Age), Stature-for-age (St/Age) and Weight-for-stature (Wt/St) among Rajput Boys of Theog Area, Himachal Pradesh

S.D. Score Category	**Weight-for-Age**		**Stature-for-Age**		**Weight-for-Stature**	
	N	**%**	**N**	**%**	**N**	**%**
> -1 S.D. scores	41	20.3	87	43.1	91	45.0
-1 to -2 S.D. scores	91	45.0	72	35.6	80	39.6
-2 to -3 S.D. scores	63	31.2	37	18.3	29	14.4
< -3 S.D. scores	7.0	3.5	6	3.0	2	1.0

nourished according to weight-for-age, 21.3% according to stature-for-age and 15.4% according to weight-for-stature.

As far as the nutritional status is concerned, the percentage of under-weight boys (low weight-for-age) was the highest. The incidence of stunting (low stature-for-age) was more as compared to wasting (low weight-for-stature) in the present sample. This indicates that more children were suffering from past under-nutrition, which adversely affected their statural growth. High prevalence of mild to moderate malnutrition among the present sample of Rajput boys could be related to their poor diet, probably on account of their poor socio-economic background. The low levels of literacy among their parents, particularly mothers, could be another contributory factor.

RELATIONSHIP OF STATURE AND WEIGHT WITH TOOTH EMERGENCE

To find out the relationship of tooth emergence with stature and weight of the boys in the present sample, partial correlation was used. Partial correlation was preferred over simple correlation in order to eliminate the effect of age on overall correlation. The values of overall partial correlation for number of emerged teeth with stature and weight were calculated to be 0.8495 and 0.8008, respectively, for Rajput boys, which were found to be statistically highly significant ($P \leq 0.001$). Thus, the number of emerged teeth in the Rajput boys of the present sample was significantly correlated with stature and weight. This indicates that the tall and heavy boys in the present sample were advanced dentally. The present results are in agreement with the findings of several other workers (Maj *et al.*, 1964; Garn *et al.*, 1965; Lee *et al.*, 1965; Billewicz and McGregor, 1975; Kaul *et al.*, 1975; Visweswara Rao, 1985, 1992; Banerjee *et al.*, 1992) who also reported dentally advanced heavy and tall children. The results of the present study suggest a relatively stronger correlation of stature, as compared to weight, with tooth emergence. This is in agreement with the findings of other workers (Vaslik, 1972; Pathak *et al.*, 1985; Visweswara Rao, 1992) who also found a stronger correlation of stature, as compared to weight, with dental emergence.

RELATIONSHIP BETWEEN NUTRITIONAL STATUS AND TOOTH EMERGENCE

Table No. 3 shows the mean number of emerged teeth among the Rajput boys, classified as per their nutritional status according to different derived anthropometric indices of nutritional status. It is clear from the table that Rajput boys classified as malnourished according to stature-for-age (those with S.D. scores of –2 S.D. or less), on the average, had less mean number of erupted permanent teeth as compared to normal children (with S.D. scores

Table No. 3

Mean Number of Emerged Teeth among Rajputs Boys of Theog Area Classified in Different Categories of Nutritional Status, according to Stature-for-age, Weight-for-age and Weight-for-stature

Nutritional Status (S.D. Score Category)	Age Group (Years) 6	7	8	9	10	11	12	13
STATURE-FOR-AGE								
> **-1 S.D. scores (Normal)**	0.6	9.2	10.4	13.7	15.2	21.2	21.0	24.7
-1 to -2 S.D. score **(Mildly malnourished)**	1.2	3.5	9.6	12.8	15.0	18.0	20.2	23.4
-2 to -3 S.D. scores **(Malnourished)**	2.4	5.5	7.0	13.0	10.0	19.5	20.0	19.2
< -3 S.D. scores **(Severely malnourished)**	-	1.3	-	-	9.0	13.0	-	21.0
WEIGHT-FOR-AGE								
> **-1 S.D. scores (Normal)**	0.7	9.5	10.5	16.0	16.7	19.2	23.5	26.0
-1 to -2 S.D. score **(Mildly malnourished)**	1.0	8.4	10.2	12.0	14.2	18.8	21.6	24.8
-2 to -3 S.D. scores **(Malnourished)**	1.4	3.2	8.2	13.0	11.6	15.8	18.1	24.6
< -3 S.D. scores **(Severely malnourished)**	2.3	2.0	-	-	-	-	-	-
WEIGHT-FOR-STATURE								
> **-1 S.D. scores (Normal)**	1.75	7.1	10.2	13.4	14.8	18	21.5	24.6
-1 to -2 S.D. score **(Mildly malnourished)**	1.7	6.3	9.6	13.5	14.4	19.4	20.2	22
-2 to -3 S.D. scores **(Malnourished)**	1.2	7.0	7.5	12.6	13.0	16.4	18	15
< -3 S.D. scores **(Severely malnourished)**	-	-	12	-	-	-	-	-

of –1 S.D. and over) in all age groups. The mean number of erupted permanent teeth among the mildly malnourished boys (with S.D scores between -1 and -2) were also less than normal boys in almost all age groups. The mean number of erupted permanent teeth among the underweight boys of the present sample (those with S.D. scores of –2 S.D. or less as per weight-for-age) were also less than that of normal children in nearly all the age categories. With a couple of exceptions, the mean number of erupted permanent teeth of the boys suffering from wasting (those with S.D. scores of –2 S.D. or less according to weight-for-stature) was also lower than that of the normal boys.

Considering all the three anthropometric indicators of nutritional status together, it is clear that the malnourished Rajput boys in the present sample had less mean number of emerged teeth than normal boys. Even those with mild under-nutrition had less number of permanent teeth than normal boys. The results of the present study thus indicate that under-nutrition, even mild under-nutrition, can have an adverse influence on permanent tooth emergence. The findings of the present investigation are consistent with those of several previous reports, which have also shown that poor nutritional status may have a negative effect on tooth emergence (Billewicz and McGregor, 1975; Niswander, 1963; Triratana *et al.*, 1990; Ulizaszek, 1996). However, the issue needs to be further investigated on larger samples from different parts of India.

BIBLIOGRAPHY

Alvarez, J.O. 1995. 'Nutrition, tooth development, and dental caries'. *American Journal of Clinical Nutrition.* Vol. LXI (Supplement), pp. 410S-416S.

Alvarez, J.O. and J.M. Navia. 1989. 'Nutritional status, tooth eruption, and dental caries: a review'. *American Journal of Clinical Nutrition.* Vol. IL, pp. 417-426.

Bailit, Howard L. and B. Sung. 1968. 'Maternal effects on the dentition'. *Archives of Oral Biology.* Vol. XIII, pp. 155-162.

Banerjee, P., Arup R. Banerjee and A.R. Banerjee. 1992. 'Eruption of permanent teeth in Bengalee school children'. *Journal of Human Ecology (Special Issue).* Vol. II, pp. 31-36.

Billewicz, W.Z. and I.A. McGregor. 1995. 'Eruption of permanent teeth in West African (Gambian) children in relation to age, sex and physique'. *Annals of Human Biology.* Vol. II, pp. 117-128.

Briend, A. and S. Zimicki. 1986. 'Validation or arm circumference as an indicator of risk of death in one to four year old children'. *Nutrition Research.* Vol. VI, pp. 249-261.

Census of India. 2001. *Provisional Population Totals, Himachal Pradesh, Paper-1 of 2001.* Director of Census Operations. Himachal Pradesh.

Chen, L.C., A.K.M.A. Chowdhury and S.L. Huffman. 1980. 'Anthropometric assessment of energy-protein malnutrition and subsequent risk of mortality among preschool age children'. *American Journal of Clinical Nutrition.* Vol. 33, pp. 1836-1845.

Fauveau, V., A. Briend, J. Chakraborty and A.M. Sarder. 1990. 'The contribution of severe malnutrition to child mortality in rural Bangladesh: Implications for targeting nutritional interventions'. *Food and Nutrition Bulletin.* Vol. XII, pp. 215-219.

Garenne, M., B. Maire, O. Fontaine, K. Dieng and A. Briend. 2000. 'Risques de dé 'cés associés a' différents états nutritionnels chez l'enfant d'âge préscolaire'. *Etudes du CEPED n_ 17,* 192 p. CEPED, Paris.

Garn, Stanley M., Arthur B. Lewis and Rose S. Kerewsky. 1965. 'Genetic, Nutritional, and Maturational Correlates of Dental Development'. *Journal of Dental Research.* Vol. XLIV, pp. 228-242.

Gragnolati, Michele, Meera Shekar, Monica Das Gupta, Caryn Bredenkamp and Yi-Kyoung Lee. 2005. *India's Undernourished Children: A Call for Reform and Action.* Health, Nutrition and Population (HNP) Discussion Paper. Washington DC: The World Bank.

Grantham-McGregor, S.M. and C.C. Ani. 2001. 'Undernutrition and mental development', in J.D. Fernstrom, R. Uauy and P. Arroyo (eds.), *Nestlè Nutrition Workshop Series Clinical and Performance Program.* Vol. 5, Nestlè Limited, Basel. pp. 1-18.

Hamill, P.V., T.A. Drzd, C.L. Johnson, R.V. Reed, A.F. Roche and W.M. Moore. 1979. 'Physical Growth: National Center for Health Statistics Percentiles'. *American Journal of Clinical Nutrition.* Vol. XXXII, pp. 607-629.

International Food Policy Research Institute. 2008. *Accelerating Progress toward Reducing Child Malnutrition in India: A Concept for Action.* Washington D.C.: International Food Policy Research Institute.

Johnston, F.E. 1981. 'Anthropometry and nutritional status', in *Assessing Changing Food Consumption Patterns.* Committee on Food Consumption Patterns. Washington: National Academy Press, pp. 252-264.

Kaul, Samvit, Sushma Saini and Bindu Saxena. 1975. 'Emergence of permanent teeth in school children in Chandigarh, India'. *Archives of Oral Biology.* Vol. XX, pp. 587-593.

Kaul, Vinod and Surya Prakash. 1982. 'Eruption of permanent teeth in Jats of Haryana', in J. Borms, R. Hauspie, A. Sand, C. Susanne and M. Hebbelinck (eds.), *Human Growth and Development.* New York: Plenum Press, pp. 243-352.

Lee, M.M.C., W.D. Low and K.S.F. Chang. 1965. 'Eruption of permanent dentition of Southern Chinese children in Hong Kong'. *Archives of Oral Biology.* Vol. X, pp. 849-861.

Maj, J., S. Bassani, G. Menini and O. Zanini. 1964. 'Studies on the eruption of permanent teeth in children with normal occlusion and malocclusion'. *Transactions of the European Orthodontic Society.* Vol. XL, pp. 107-130.

Mason, J.B., P. Musgrove, J.P. Habicht. 2003. *At least one-third of poor countries' disease burden is due to malnutrition.* Bethesda, MD: Disease Control Priorities Project. Fogarty International Center, National Institutes of Health (Working paper no. 1).

May, M. and M.V. Montagu. 2008. 'Malnutrition and food security: creating 'appropriate GM' for the poor farmer'. *Real Health News.* No. 9, pp. 40-43.

Mayhall, J.T., P.L. Belier and M.F. Mayhall. 1978. 'Canadian Eskimo permanent tooth emergence timing'. *American Journal of Physical Anthropology*. Vol. XLIX, pp. 211-216.

McLaren, D.S., E. Shirajian and S. Loshkajian. 1969. 'Short term prognosis in protein-calorie malnutrition'. *American Journal of Clinical Nutrition.* Vol. XXII, pp. 863-870.

Mellanby, M. 1928. 'The influence of diet on the structure of teeth'. *Physiological Reviews.* Vol. VIII, pp. 545-77.

Murray, C.J. and A.D. Lopez. 1997. 'Global mortality, disability, and the contribution of risk factors: Global Burden of Disease Study'. *Lancet.* Vol. CCCXLIX, pp. 1436-1442.

Niswander, J.D. 1963. 'Effects of heredity and environment on development of dentition'. *Journal of Dental Research.* Vol. XLII, p. 1288.

Pathak, Raj K., Sushma Jaiswal and Samvit Kaul. 1985. 'Association of permanent dental maturity status with stature and weight in Punjabi children of Chandigarh, India', in V. Rami Reddy (ed.), *Dental Anthropology: Applications and Methods*. New Delhi: Inter-India Publications, pp. 49-53.

Pelletier, D.L. 1994. 'The relationship between child anthropometry and mortality in developing countries: implications for policy, programs and future research'. *Journal of Nutrition.* Vol. CXXIV (10 Supplement), pp. 2047-2081.

Pelletier, D.L., E.A. Frongillo, D.G. Schroeder and J.P. Habicht. 1995. 'The effects of malnutrition on child mortality in developing countries'. *Bulletin World Health Organization*. Vol. LXXXIII, pp. 443-448.

Ramachandran, P. 2007. 'Poverty Nutrition Linkages'. *Indian Journal of Medical Research.* CXXVI, pp. 249-261.

Seow, W.K. 1996. 'A study of the development of the permanent dentition in very low birth weight children'. *Pediatric Dentistry.* XVIII (No. 5), pp. 379-384.

Schour, I., M.M. Hoffman and M.C. Smith. 1941. 'Changes in the incisor teeth of albino rates in vitamin: A deficiency and the effects of replacement therapy'. *The American Journal of Pathology*. Vol. XVII, pp. 529-62.

Shaw, James H. and Derrick Gniffiths. 1963. 'Dental abnormalities in rats attributable to protein deficiency during reproduction'. *Journal of Nutrition*. Vol. LXXX, pp. 123-41.

Tanner, J.M. and R.H. Whitehouse. 1966. *Notes on the use of height and weight standard charts*. London: Creasey, Hertford.

Triratana, T., Hemindra and C. Kiatiparjuk. 1990. 'Eruption of permanent teeth in malnutrition children'. *The Journal of the Dental Association of Thailand.* Vol. XL (No. 3), pp. 100-108. (in Thai)

Ulizaszek, S.J. 1996. 'Age of eruption of deciduous dentition of Anga children, Papua New Guinea'. *Annals of Human Biology.* Vol. XXIII (No. 6), pp. 495-499.

Vaslik, J.A. 1975. 'Changes in eruption of the first permanent teeth'. *Scripta Medica.* Vol. XLVIII, pp. 191-194.

Vella, V., A. Tomkins, J. Ndiku, T. Marshal and I. Cortinovis. 1994. 'Anthropometry as a predictor for mortality among Ugandan children, allowing for socioeconomic variables'. *European Journal of Clinical Nutrition*. Vol. XLVIII, pp. 189-97.

Visweswara Rao, K. 1985. 'Nutritional status, deciduous dental eruption and age assessment', in V. Rami Reddy (ed.), *Dental Anthropology: Applications and Methods*. Inter-India Publications, New Delhi, pp. 3-11.

——. 1992. 'Differential growth and dental eruption in Indian children'. *Journal of Human Ecology (Special Issue)*. Vol. II, pp. 25-30.

Waterlow, J.C., R. Buzina, W. Keller, J.M. Lane, M.Z. Nichaman and J.M. Tanner. 1977. 'The presentation and use of height and weight data for comparing the nutritional status of groups of children under the age of 10 years'. *Bulletin of the World Health Organization*. Vol. LV, pp. 489-498.

Weiner, J.S. and J.A. Lourie. 1981. *Practical Human Biology*. London: Academic Press.

World Health Organization. 1981. *Development of indicators for monitoring progress towards health for all by the year 2000*. Health for All Series No. 4, Geneva: World Health Organization.

World Health Organization. 1983. *Measuring Change in Nutritional Status*. Geneva: World Health Organization.

World Health Organization. 1986. 'Use and interpretation of anthropometric indicators of nutritional status'. *Bulletin of the World Health Organization*. Vol. LXIV, pp. 929-941.

10

REPRODUCTIVE HEALTH: PERCEPTION AND PRACTICES AMONG THE TRIBAL WOMEN OF GUJARAT

Ratnawali

INTRODUCTION

Notion of health and illness is influenced by the complex interaction between individuals, their socio-cultural settings and physical environment (Hollingshead, A. and Redlich, F. 1958). Health, thus, cannot be seen in isolation of its social context. Apart from epidemiological consideration, it essentially is a social phenomenon that exists within the realm of culture and its niche. The interplay of cultural dimensions in understanding the deviation in normal functioning of body provides a tool to deal with the malaise within the socio-ecological framework. It entails a comprehension derived from the observable or non-observable facts that underlie the cause and try to treat it with the available resources and wisdom lying with the community. In this paper some such perception and beliefs of the tribal women with regard to their reproductive health have been brought out, which besides providing an insight into the cultural dimension of health practices, also reveal limitations of modern health facility in providing credible and affordable services. The traditional practices are sometimes relied as the only available option to the people. Data have been collected through focus group-discussion with women, who are in their reproductive age (15-48 years), from Dang as well as from Dharampur and Kaparada talukas of the Valsad District. These are numerically dominant tribal areas. The paper is divided into four sections. The first section is the context of the study; the second section deals with the profile of the study area. Beliefs and practices are dealt with in the third section and the last one presents the changing trend and conclusion.

I

All cultures have their own concept of health, hygiene, disease and cure which vary considerably across the globe. Understanding of disease and health problem is largely contextual to the cultural settings. The social construction of illness and their internalisation by the cultural group determine the mode of treatment which in turn directs the treatment seeking behaviour. Thus, disease and its perception play an important role in health seeking behaviour within a culture. It is beyond doubt that construction of such perception shapes from the human experiences and their interaction with the natural world. It is also true that specialised knowledge to these diseases and their cure is believed to be possessed by few in the community who are expected to provide relief from the sufferings. Several studies have shown the influence of cultural beliefs on the health service seeking behaviours. Carstairs (1955) had brought out how difference of perception in modern and traditional system hinders the medical care supposed to be provided by the doctor in his study on Rajasthan village. Traditional system that entails a close relationship between provider and subject, many a times does not have the same closeness in modern system as is highlighted in the study by Marriot (1955) in Kishengarhi, a village in Uttar Pradesh. Role and expectation of the two systems act as contrast to the social setting in the above study, impeding the use of modern system by the villagers. Hasan (1967) in a holistic study in "The Cultural Frontier of Health in Village India" looked into all aspects of health in its social milieu including interpersonal relationship, ecological surroundings and analysed their role in treatment seeking behaviour. Madan (1969) and Kakar (1972, 1973 and 1977) have explained that rural people attributed diseases to certain external conditions which involve natural as well as supernatural powers acting in conjunction with certain internal conditions of the body. Similarly, strong influence of beliefs, attitude and value make an impact on the interpretation of the symptoms of disease and the techniques for their treatment as reported by Chopra (1984) in four villages of Haryana. He also observed the limited capacity of health centres in catering to the demand of health care of substantial population of these villages. Preventive and curative practices of health among tribal population have been widely studied (Basu *et al.*, 1990; Mahapatra, 1990; Basu, A., 1990). Concept of disease, notion towards treatment and cause-and-effect relationship are well documented among Oraons, Kanjaras, Koyas, Enadi, Baigas and Khonds by Surendra (1990). These studies have brought to the focus various health practices prevailing in the tribal societies to deal with their illness.

Reproduction being an innate desire to further the propagation of species, it always finds attention in all human societies with indigenous practices for safe motherhood. The physiological changes in the women during pregnancy and the process of child-birth always hold them in awe with mystic understandings. Not surprisingly, frequent death of mother and child and complication arising out of various internal and external factors lead to the traditional practices and belief systems which beside devising curative measures also laid emphasis on dos and don'ts to deal with the situation. This is also true about other gynaecological problems that women suffer from. The cross-cultural study by Mark Nichter (1989) examined such issues with ethno-physiological and ethno-anatomical perspective. He has documented perceptions regarding fertility, folk dietetics during pregnancy, fertility regulation and cultural interpretation of malnutrition that give insight into people's reasoning about their practices. Kar (1993) studied the reproductive health behaviour among Nocte women in Arunanchal Pradesh and found out observance of taboos to ensure the well being of expecting mother and her foetus. Use of herbal medicine and services of medicine man in family planning practices have been reported from Nicobar Island by Durai and Sahay (1997). Thus, beliefs regarding diet, taboos, rituals and herbs all make a composite health system that deal with the safe delivery. The efficacy of this age-old tradition is greatly valued and individuals imbibing these values in course of enculturation seldom drift to other systems.

Any health interventions that are directed to provide medical services to population require deep insight into their belief system to mould their perception towards modern health services. In the absence of such initiative all plan, policy and infrastructure will have little impact on the target population. Evidences to these limitations can be found in various health schemes taken up by the government from time to time. Mother and child health have received attention since Independence, but the results have hardly matched the expectations as envisaged during plan periods since then. The rechristening of MCH programme, which was designed to look after the mother and child health to RCH (Reproductive and Child Health) after ICPD conference (1994), was to enhance the overall health of the mother and child along with the critical health concerns of the mother related to reproduction. However, goal to this effect is much debated (Visaria and Ramachandran, 1999; Chatterji, M., 1999; Jejeebhoy, S.J., 2000; Kumar, R., 2002) and it has been brought to the focus that women's health needs still suffer from lack of commitment from the public authority consequently, from the pursuance of the age-old belief system that are not capable enough to offer her relief from gynaecological problems. This is more so with the tribal communities where their poor health status is invariably linked with their reluctance to accept medical assistance. However, an effort to find out the reason behind them is seldom attempted. Exploration into cultural beliefs and practices opens up a cognitive world of the community, which helps in initiating relevant medical interventions.

II

Gujarat has a sizeable tribal population that accounts for 14.9 per cent of the total population and is an abode to twenty-nine tribal communities distributed over eleven districts located along the eastern border of the state. The highest concentration of tribals is found in South Gujarat with districts of Surat, Navasari, Valsad, Bharuch, Narmada and Dang housing more than 60 per cent of the total tribal population. The socio-economic indicators for tribals are relatively poor in comparison to the general population of the state. Literacy rates among tribals are eighteen points lower than the state's average. It is 51.78 among them whereas 69.1 for the state as a whole. The ecological degradation due to various development projects in their habitat, apart from adversely impacting their livelihood has also recorded highest incidence of poverty against them (Hirway 2004). Some of the health indicators for the districts with sizeable tribal population are presented below:

Table No. 1
Selected Health Indicators for Some of the Districts with Sizeable Tribal Population

Districts	**Crude Birth Rate (1991)**	**Crude Death Rate (1989-90)**	**Total Fertility Rate (1991)**	**Infant Mortality Rate (1991)**	**Child Mortality Rate (under 5) (1991)**	**Morbidity Rate (92-94) Per 1000**
Bharuch	30.28	9.71	3.63	49	86	67.14
Dangs	39.19	9.71	5.04	87	112	83.77
Panchmahals	34.37	6.49	4.70	79	111	97.85
Sabarkantha	34.72	6.02	4.26	75	104	34.88
Surat	30.64	6.63	3.45	60	88	123.60
Valsad	28.96	7.83	3.43	55	83	40.39
Gujarat	**33.56**	**8.80**	**4.20**	**78**	**101**	**55.00**

Source: Gujarat human development report 2004, Table 5.4 and 5.10.

Apart from these indicators, malnourishment among children is highest in tribal communities. In the age group of less than 3 years, 24.1 per cent of Schedule Tribe children are severely underweight whereas 29.2 and 3.4 per cent belong to severely stunted and wasted category. The same parameters for Gujarat as a whole are 16.2, 23.3 and 2.4 respectively (IIPS and MEASURES DHS+ (2001: 170).

Women in general and tribal women in particular are more vulnerable to health-related problems. The difficult living conditions and improvised health facilities further lowers their chances of optimum good health. It should be noted that tribal habitats are located far and wide in the hilly terrains. Aside from improvised health facilities, access to the same is also constrained by poorly developed infrastructure like road and transportation. Malnutrition, anaemia and reproductive health-related problems are recorded in higher proportion among these populations. A comparative account of some of the health indicators are given in Table No. 2.

Table No. 2
Health Status of Schedule Tribe Women of Gujarat

Health Indicators	**Schedule Tribe Women**		**Gujarat (all women)**	
	1993	**1998**	**1993**	**1998**
Body Mass Index *	NA	19.0	NA	20.7
% with BMI below 18.5 kg/M2 *	NA	55.0	NA	37.0
% of Anaemia	NA	55.5	NA	46.3
% with no Ante Natal Check-up	31.5	22.2	23.8	13.6
% with no Tetanus Toxoids vaccination	42.0	27.0	30.2	17.6
Iron folic (%)	62.4	86.9	69.3	85.4
Delivery by TBA (%)	68.2	60.6	45.5	42.4
Use of any contraceptive method (%)	47.3	57.1	49.3	59.0
Symptoms of urinary tract infection in ever married women (%)	NA	11.4	NA	12.3
Any reproductive health problem (%)	NA	29.1	NA	28.6

Source: NFHS-I and II, * IIPS and MEASURES DHS+ (2001)
Note: NA – not available

Though, in general, the health status of tribals, particularly the reproductive health of women have been lower than the general population, it has registered improvement over the years as the two-time series data (NFHS-I and II) reveal. It also speaks of improved effort on part of health workers serving in these areas. However, a significant proportion still remains outside the formal health services and is totally dependent upon their traditional practices for their health requirements. Even those who access formal health services, also seek the services of traditional healers. The traditional practices, though, may not be scientifically sound nonetheless; they serve the function of some psychological relief resulting out of their faith in the system.

In the above context, the present study report some findings from the Valsad and Dang districts that comprise 54.35 per cent and 93.8 per cent of the tribal population respectively. As I have mentioned earlier, Dharampur and Kaparada of Valsad district have been chosen for their preponderance of tribal population. Dang is taken as a whole for it is not differentiated into talukas. The demographic profile of the study area is as below.

Table No. 3
Selected Demographic Features of Valsad and Dang Districts

Name of District/ Taluka	Total Population	% of ST Population	Sex Ratio (general)	Sex Ratio (ST)	% Literacy			Growth Rate 1991-2001
					Total	Male	Female	
Dharampur	180197	91.8	984	994	48.59	58.25	38.81	22.85
Kaparada	203009	93.6	986	993	34.48	43.18	25.66	28.88
Valsad	1410680	54.8	920	999	69.41	78.10	59.92	29.66
Dang	186,729	93.8	987	995	59.7	70.70	48.50	29.58

Source: Census 2001 (provisional), Gujarat.

The district of Dang is completely rural whereas that of Valsad is comprised of 77 per cent rural population. Taluka-wise, Dharampur has 88 per cent rural population while all of Kaparada is rural. Both the talukas of Valsad have more than 90 per cent tribal population and are low on literacy figures compared to Dang. Female literacy is further low and is particularly poorer in Kaparada. Sex ratio is, however, far better in the areas which are exclusively dominated by tribal population. Hence, Dharampur, Kaparada and Dang have higher sex ratios than Valsad implying more egalitarian attitude of tribal community towards women. For the purpose of study, six villages each from Dharampur, Kaparada and Dang were selected on the basis of their distance from the taluka headquarters (Table No. 4). Focus group discussion and in-depth interview were conducted in each village for the purpose of the study. Altogether 226 women from Valsad and 88 women from Dang participated in focus group discussion. The average age of the women were 30 years in Valsad and 28.6 years in Dang belonging to Kunkani, Bhil, Varli, Dhodhia and Kolgha tribes.

Table No. 4
Taluka/District-wise List of Selected Villages

Taluka/District	Name of Selected Villages
Dharampur	Ojharda, Vavar, Malghar, Suliya, Khadagavan and Ambheti
Kaparada	Nani Dholdungri, Tamchadi, Sidumber, Vaghvad, Vankhas and Singarmal
Dang	Mulchod, Natkatia Hanvant, Nani Dabdar, Malegam, Godaria, Iskhandi

III

Perceptions are understandings or impressions about a particular event, object or person whereas belief is the social product of individual perception as well as group experience of any social phenomena (Pokarna, 1994). The cause-and-effect relationship as viewed by the group is credited, acknowledged and incorporated into the belief system, which is perpetuated and internalised through the process of socialisation. This institutionalised system of belief about disease and illness directs the course of treatment seeking behaviour. Rationality in folk belief system may not have modern standard but as it has evolved over years of experience, it is pursued with much sincerity especially in tribal communities. This being the case rituals, observance of dietary restriction and magical as well as medicinal treatment has been the part of tribal health systems in general and reproductive health in particular for long.

Women in the above-mentioned tribal areas figured out menstruation (*masik*), leucorrhoea (*safed pani*), pregnancy (*mahina*) related problems, delivery (*suvavad*), abortion (*Garbh padi jaye, Kasuvavad*) and infertility as their main problems regarding reproductive health. Other problems like itching (*khajevat*) and foul-smelling discharge from genitals are also reported.

Menses is perceived as impurity within body and is attached with the notion of pollution. Prohibition is observed in cooking, touching drinking water and visiting sacred places. Such observance is less reported from Dang. Pain, giddiness, nausea and weakness during menstruation is taken as normal and no medicine is sought for the same. However, in case of prolonged and heavy bleeding, women approach the *Bhagat* (traditional medicine man) for treatment. Less bleeding with accompanying pain is perceived as accumulation of blood clots that are caused by some gaseous force obstructing the free flow. In such cases also, medicines are taken from *Bhagat*.

Leucorrhoea is widely reported in all focus group discussions but treatment is seldom sought. In Dang it is perceived to be caused by swelling in the uterus. Only in case of heavy discharge, fever and weakness, is the help of *Bhagat* sought. Sometimes nurses are also contacted for medicines if problem persists and affects routine working of the women.

Notions and beliefs regarding child-birth are extensive and are observed with sincerity. The problems that are encountered during pregnancy are night blindness, oedema, abdominal pain, weakness and vomiting which are perceived as common problems and no medication is sought for them. Bleeding in this period is, however, taken seriously and the *Bhagat* is contacted at the earliest. It is believed in Dang that such problems are caused by bad air impregnated with evil spirits and the *Bhagat* being knowledgeable about them is the only person capable to cure it. Such beliefs find little mention in the villages of Dharampur. At Dang, pregnant women are not supposed to venture out in the evening for fear of harms supposed to be inflicted upon by evil ghosts. Restriction on certain diet is commonly observed across the villages in both the districts. Thus, curd and oily things are prohibited as it is believed that being sticky in nature, they make the child stuck in the womb. It will ultimately lead to difficulty in delivery. Banana, papaya, peanuts and black grams are also avoided for their heating attributes fearing miscarriage. Similarly, any type of allopathic medicine is believed to cause heat and is refrained from consuming to save the baby from its impact. Eating in small plate is advised for pregnant women assuming that it will not let the placenta grow big. The bigger the size of the placenta, the more it would be difficult to deliver, leading to increased risk to the life of woman and child. It was extensively believed in Valsad that eating in hiding would result in hiding of placenta within the body and the woman will face complications at the time of delivery.

Problems like prolonged labour during delivery, heavy bleeding prior to delivery and fluid discharge long before the birth of the child (*Sukhi Suvavad*) are of great concern to the women. As is believed in the Dang, eclampsia or convulsions (*khainch*) during or after the delivery is caused by the spirits who managed to take possession over women (*Bhoot Bharaya*). *Bhagats* are called in to cure the problem. Delivery is preferred to be done at home with the help of *Dais* (traditional birth attendant or *Dai*) who in most of the villages were trained but services of untrained *Dais* are also not uncommon and are reported from some of the villages of Kaparada and

Dang. Elderly women sometimes assist *Dais* in delivery. In case of prolonged pain, some herbs are placed beneath the back of the women during labour to facilitate easy and safe delivery. Practices like stuffing bunch of hair in the mouth of the woman or giving her a hard smack are also observed to hasten the delivery of the baby. These are done with the understanding that women under shock and desperation will deliver fast. However, these practices are not routinely followed, and women are preferred to be taken to hospital if situation and resources allow. In Valsad, expecting mothers diagnosed with transverse position of child during advanced pregnancy are referred to hospital, but in Dang, women stated that such cases are handled by *Dais* who are skilled to correct the position of the foetus through massage. One may wonder whether such practices are really effective or are one of the contributing factors in the death of baby and sometimes mother, since such cases were more frequently reported from the Dang.

Similarly, stillbirths or death of the baby soon after delivery due to entangled placenta are perceived as handiwork of ghosts and other malevolent spirits who target pregnant women venturing out at night. These spirits entangle the placenta around the baby's neck, thus strangulating it within the womb. Such beliefs are common among women in Dang and Kaparada.

Post-delivery complications like backache and weakness are considered normal. Drugs for general weaknesses or other problems are avoided as they are considered hot and would dry lactation. In some of the villages of Dang, liquor is given to the women to ease the pain soon after the delivery whereas in other places *Amba Haldi* (a herb) is administered to heal the injury. The mother not allowed to go out for some days after delivery as she is perceived to be low on inner strength *(Bhini Stri)*, hence susceptible to fall easy prey to evil spirits. Pollution after delivery is observed for at least five days across the villages but in some places it is extended to one-and-a-quarter month. Rice gruel or *Dhanchlu* is given at least for the first five days after delivery. Afterwards, pulse, rice, dry loaf, *bajara* or *jowar* loaf without its hard fringes are given as these are easily digestible. Sour, oily and spicy food, black gram, onion, garlic and brinjal are avoided for they cause sepsis in the umbilicus of the child. Infants are given breast feed soon after birth but in some areas of Dang, the first watery fluid is discarded in the belief that it may cause heat inside the baby which may result diarrhoea.

During FGDs, miscarriage or spontaneous abortion was found to be of common occurrence across the villages. Along with the spirit's interference, causation is also perceived to have physiological dimensions. Impact of evil spirits is a widely held belief and is differentiated into direct or indirect influence of the same. The mother gets into direct influence when she comes in contact of air, polluted with bad spirit resulting in sudden miscarriage. In such cases, precautionary measures in the subsequent pregnancy are observed. It includes tying of sacred thread by *Bhagat* or wearing fetishes to ward off evil spirits. The indirect influence is attributed to *Bhamra* (special form of hair twirl at the back of the head or neck) seen in the preceding child and is observed as bad omen. The evils, it is believed, see through *Bhamra* and extend their impact on carrying mother resulting in loss of pregnancy. Under such circumstances, the child is branded and some rituals are followed to ward off evil influences. Branding is called as *Daam* and is practiced in both the districts. The physiological problem as discerned by the *Bhagat* is believed to be caused by impurity within the uterus (*kotha*) that expels the child out. To cleanse it, the *Bhagat* makes a slit on the abdomen and pack it with herbs. It is termed as *Faans* and is practised across the villages but more frequently in the interior villages of both the districts. Doctors are consulted only after the treatment by *Bhagat* fails.

Induced abortion is perceived as sin and is not accepted openly. During in-depth interview, it was revealed that women do resort to such practices to get rid of unwanted pregnancies. Medicines from *Bhagats* or self-administered home remedies like herbal concoctions, papaya seeds, hot fomentations are tried to end pregnancies. However, there has been a realisation that these methods are not really effective and can result in disfigured child. Therefore, secret visit is made to the health centres to abort the pregnancy. However, women in need look for a knowledgeable person in and around to avoid complex procedures of public facility and high expenditure of private clinics.

Barrenness or infertility is particularly perceived as women-specific and may even pave the ground for divorce. To cure infertility, *Daam* and *Faans* are widely practised linking its causation to spirits as well as to physiological ones. Along with above practises, vows, fetishes and observance of certain rituals are also followed to appease God. Though, the *Bhagat* is considered more capable in curing infertility with his rituals and other practices, nevertheless, women do pay visits to doctors to overcome the problem.

Issues related to STDs or RTIs were refrained from discussion as women show ignorance about these matters in their village. However, in two FGDs a passing reference was made about the visits paid to the local *Bhagat* for complaints relating to foul discharge or blisters on the genitals. Since these problems are regarded as physiological ones, a visit to the doctor is considered necessary in addition to the herbal medicines given by the *Bhagat*.

Contraceptive practices are not in vogue as is the knowledge of the same among women. Despite little information and low awareness about various methods, side-effects like vomiting, abdominal pain and heavy menstrual flow are cited as problems by those who were using it. In general, these problems dissuade them from adopting any method other than sterilisation. Experience of ineffectiveness of the above methods in some cases also deters women to use them. Thus, after having 3-4 children with at least one son, sterilisation is preferred to limit the family size. Easy accessibility of sterilisation with ANMs facilitating their travelling and other medicinal expenses to the health centres and back home also makes it popular among women. Male sterilisation is not the norm as it is perceived to foster weaknesses in them. However, in Dang some women affirmed that their husbands had undergone vasectomy to limit the family size. Practices of any traditional contraceptive methods, however, were not reported from the study area.

IV

The extensive beliefs and observation of rituals related with child-birth specifically, can be interpreted as a pointer towards insecurity that a community experiences in efficiently dealing with the inherent dangers that are likely to unfold during child-birth. Continuance of age-old traditional practices not only calls for investigation into the efficacy of the system but also an examination of the other options available to the communities. In the above context, the provision of public health facilities appears to be very limited. There is a clear lack of proper infrastructure, inadequate manpower and commitment on part of the public health functionaries. The sub-centres are poorly maintained and are altogether absent in some cases. The ANMs run them from Anganwadi centres or they function only for immunising children. It also deters ANM to stay in the village, considerably constraining the heath services which are already meagre and limited to external abdominal examination, dispensing of iron folic tablets and immunisation only. Provision of services like delivery and postpartum checkups are nearly absent whereas cases like complicated delivery, pneumonia in infants and other emergency situations suffer for lack of proper transportation and skilled personnel. The health personnel's general scorn towards traditional practices and little willingness to reach out to women further reduces the acceptance of formal health services among women in the interior villages. These villages being situated in the hilly terrain also lack in all-weather roads, deterring their access to medical facility especially during monsoon. The problem is further compounded by the uncertainty of drugs as well as doctors. Thus, the traditional practices like *Daam* and *Faans* that many a times lead to sepsis, continued to be observed for want of adequate health provisions and negligible dissemination of health awareness among the women. It is particularly more so with the interior villages of Dang and Kaparada where greater reliance is placed on traditional health practices than those villages that have access to some kind of health facility.

However, a subtle change has been observed in the attitude of women. A gradual awareness in health care utilisation is apparent from 1993 to 1998 NFHS data. The field study also recorded this change. At the village level, three factors seem to influence them. These are education, accessibility of health services and greater contact with the outside world by means of improved roads and communication.

The level of education though has not reached the desired stage; nevertheless even the less literate women do seek the services of health professionals wherever possible. This is apparent in the attitude of those young women who have received education even up to upper primary school. They understand the benefits of modern health practices and seek services like antenatal check-up and follow-ups advised by the doctors. Sometimes they also act as change agents and influence their peer groups. In the focus group discussion, some of these women categorically stated that *Bhagat's* treatment is not beneficial in most of the cases, but their views came out strongly only on few occasions. Though these views were not refuted, they were not shared by all either.

The availability of health facilities and the attitude of health personnel play a big role in the acceptance of modern health services in the communities. Thus, villages which were near the health facilities were definitely utilising the formal health services more than the distant ones. The ANM and doctors in these villages were sometimes consulted even prior to the visit to the *Bhagat*. Inadequate availability of health services automatically transfers dependence upon *Bhagat* and traditional services. Better roads and communication facilities not only have attracted the private practitioner to these villages but also offered various other options to the women as well. However, such facilities were hardly offering any service to women-specific problems.

There has been a growing demand for modern health care facilities as women have conceded limited success to the *Bhagat* in treating complicated cases. A general discontent with the inadequate facility and quality of health service was palpable. However, delay in relief from allopathic medicine and whatever little success the *Bhagat* might have by whatsoever means, gave him an edge over modern health services. Many a times women have reverted back to the *Bhagat* when formal medicines have not been effective in giving her relief. It calls for a careful assessment about the institution of *Bhagat* and their herbal medicines that can no longer be dismissed as fake. It needs to be researched and promoted if found effective. The rituals accompanying the traditional medicines would have a psychological assurance to the service seekers which need not be challenged unless they cause physical harm to women. It must be noted that even the modern medical system does not offer panacea for all illnesses.

The observation by Leiben (1973) that "despite the increasing utilisation of modern medicine in developing areas with consequent reduction in morbidity and mortality, traditional medical systems still persist and exert a significant influence on the state of the health and on medical decision and out comes in developing societies" still seems to validate the present situation. While adequate formal health facility is the need of the hour, importance of traditional health system can no longer be ignored. It requires a thorough look into the curative value of the latter by weaning away harmful practices. Promotion of indigenous system, besides being cost-effective, is also more sustainable and acceptable to the people who have been practising them for centuries.

BIBLIOGRAPHY

Basu, A. 1990. 'Anthropological approach to tribal health', in Ashis Bose *et al.* (eds.), *Tribal Demography and Development in North-East India*. Delhi: B.R. Publishing Corp.

Basu, S.K., A. Hindal and G. Kshatriya. 1990. 'Determinants of health seeking behavior among the tribal population of Bastar district, Madhya Pradesh, India'. *South Asian Anthropo. II* (i): I.

Carstairs, G.M. 1955. 'Medicine and Faith in Rural Gujarat', in Benjamin D. Paul (ed.), *Health Culture and Community*. New York: Russell Sage Foundation.

Census of India (provisional), 2001, Government of India.

Chatterji, M. 1999. 'Gujarat', in Leela Visaria *et al.* (eds.), *The Community Need Based Reproductive and Child Health in India: Progress and Constrains.* Jaipur: Health Watch Trust.

Hasan, K.A. 1967. *The Cultural Frontier of Health in Village India*. New York: Manktalas.

Hirway, Indira and Darshini Mahadeviya. 2004. *Gujarat Human Development Report.* Ahmedabad: Mahatma Gandhi Labour Institute.

Hollingshead, A. and F. Redlich. 1958. *Social Class and Mental Illness.* New York: Willey.

Jejeebhoy, S.J. 2000. 'Safe Motherhood in India', in R. Ramasubban and J. Jejeebhoy Shireen (eds.), *Women's Reproductive Health in India.* Jaipur and New Delhi: Rawat Publication.

Joshi, P.C. and Anil Mahajan. 1990. *Studies in Medical Anthropology.* New Delhi: Reliance Publication.

Kakar, D.N. 1972. 'The Health Centre Doctor and Spirit Medium in a North Indian Village'. *The Eastern Antropologist* XXV, No. 3.

——. 1977. *Folk and Modern Medicine: A North Indian Case Study*. New Delhi: New Asian Publishers.

Kar, R.K. 1993. 'Reproductive Health Behaviour of the Nocte Women in Arunachal Pradesh'. *Social Change* 23(4): 40-52.

Madan, T.N. 1969. 'Who Choose Modern Medicine and Why?'. *Economic and Political Weekly* (4)37: 1475-84.

Mahapatra, S. 1990. 'Cultural pattern in health care: a view from Santal world', in B. Choudhuri (ed.), *Cultural and environmental dimensions on health.* New Delhi: Inter-India Publications, pp. 249-58.

Marriot, M. 1955. 'Western Medicine in a Village of Northern India', in B.D. Paul (ed.), *Health Culture and Community*. New York: Russell Sage Foundation.

National Family Health Survey 1993, 1998. Government of India.

Nichter, M. 1989. *Anthropology and International Health.* Netherlands: Kluwar Academic Publishers.

Pokarna, K.L. 1994. *Social Beliefs, Cultural Practices in Health and Disease*. Jaipur and New Delhi: Rawat Publication.

Visaria, Leela *et al.* 1999. 'Overview'. *The Community Need Based Reproductive and Child Health in India: Progress and Constrains.* Jaipur: Health Watch Trust.

11

CULTURAL PRACTICES AND THEIR EFFECT ON REPRODUCTIVE HEALTH AMONG THE TIWAS OF ASSAM

Rijumoni Sarma

INTRODUCTION

Health is one of the principal assets of every human being. The World Health Organization (WHO) defines health as "a state of complete physical, mental and social well being and not merely the absence of disease and infirmity". In different parts of the world, man devises numerous mechanisms for ensuring health and eliminating disease. Health and disease are indispensably related to the bio-cultural spectrum of a community in a particular environment. Both in simple and traditional societies, these traits are deeply rooted in the socio-cultural matrix. The medical history of man can only be understood through an in-depth study of the different socio-cultural systems of the human being.

In recent years, more attention is being given to the problem of rural health, particularly for the tribal and other backward groups, who are generally backward in a number of ways, but represent a sizeable population of India. Various medical facilities are given to them but very often they do not utilize the facilities available to them. It is because a large number of tribal communities are there with varied socio-cultural traditions, economy and interaction with outside world, the concept of disease and nature of treatment are likely to be different.

The cultural practices of any community have direct effect on the health of the mother as well as the child.

During pregnancy, various taboos, beliefs, special food including vitamin, minerals, etc. are given to the mother.

Each and every culture has its own concept of health and strategies for coping with diseases, which are often referred to as health culture. Various locally available herbs are used for curing diseases. In the pristine societies, people believe that evil spirits, black magic, breach of taboo, witchcraft, sorcery, etc. might lead a person to illness. Therefore, the pattern of behaviour and belief of the people relating to health and disease are to be accounted for the study of health.

THE LOCALE

In the light of the above-mentioned traits, an attempt has been made to examine the cultural practices and their effect on the reproductive health condition of the two Tiwa villages of Morigaon district, Assam. The two villages are homogeneous villages consisting 77 households in Manipur village and 65 households in Mikirgaon-Solmari village. The Tiwa is a Scheduled Tribe of Assam. The people of these two villages are mainly agriculturists. Besides these, they also practise fishing. The percentage of literate persons in the two villages is very low.

MATERIALS AND METHOD

Information was collected through anthropological techniques like observation, interviews, case-studies, etc. In order to achieve more reliability in data, reported information was cross-checked with the documents of local hospitals, primary health centres, etc.

CULTURAL PRACTICES AND ITS EFFECT ON REPRODUCTIVE HEALTH

Food habit plays an important role in the health of the people. The people of the two villages still consume their traditional food items, which are cooked with a little bit of oil. Boiled rice is the staple food of the people. Flesh of chicken, pig, goat, and other fowls are used by them. Dry fish, *khar*, *kachu*, etc. are regularly used in their food items. Besides these, *Bhedailata* (*Herpestis monnieria*) is given to a woman during post-natal period.

In the summer season, bitter dishes are prepared from the flower of *bahaka* (*Adhatada vasica*) or night jasmine (*Nyctanthes orbortristis*) and leaves of *neem* (*Melia azadirachta*) to get rid of worms. *Kaldil*, *manimuni*, etc. are used in their daily food items. These food items give them unknowingly lots of minerals, irons, protein, carbohydrate, vitamins, etc. which are necessary for the human body.

But the lower income group people do not use them regularly. The health condition of 70% pregnant women of the two villages are good. Only

30% women have some deficiencies, dysentery, vomiting, etc. The Tiwa women of these two villages generally ignored to take medical facilities. Through anthropometrical measurements, normal health is found in 85% women in Mikirgaon and 81.5% women in Manipurgaon. Still it is not sufficient. It is because of their unhygienic living condition, poor sanitation, their super-natural beliefs, low income, insufficient food gradients, etc.

Various health-related programmes are launched for the rural people. But from survey, it is found that the people are not interested to take these facilities because they could not completely come out of their supernatural beliefs. The poor communication system also plays an important role in their case.

The road systems of the two villages are not good and this creates many problems during delivery time. Small vehicles are available during daytime. However, at night these are not found. Therefore, they mainly depend on their traditional way of delivery. Elderly women of the village generally help them in such cases.

The economic condition plays an important role in this case. The economic condition of Mikirgaon is better than that of the Manipurgaon. Because of this reason, they preferred their traditional way of delivery.

"The Janani Suraksha Asani" is launched for the welfare of the rural people. But the people of these two villages are not interested in this facility. It is because of their wrong concept towards hospital delivery. Sub-health centres are present in these two villages. The nurses here perform their duties properly. They give iron tablets and folic acids to the expected woman but they do not have these regularly. The supplies are not fulfiling the requirements of the villagers. Therefore, they use locally available herbs for the cure of disease. The 'Ashas' are employed for the fulfilment of the health programme regarding pregnancy. They survey house to house and find out the pregnant women.

They do their best but the people are not interested to take the vaccines and tablets which are necessary during pregnancy.

One affect is also seen of this health programme regarding delivery. Among them, some women are again conceiving to get the money which is given by government hospitals to take care of the baby after delivery.

The women are hard workers. They not only do the household duties but also other work such as jobs in the paddy field, fishing, weaving, etc. Perhaps for this reason they have fewer problems during pregnancy and at the time of delivery. The general symptoms such as weakness and vomiting are also seen among them during the early stages of pregnancy but with the passage of time, these symptoms disappear.

BELIEFS AND PROPITIATION

The villagers believe that many evil spirits and deities can cast disease to human beings. To avoid such situations, they strictly observe some pristine taboos. During pregnancy, they are given some *bandhanies,* which are generally tied in the neck, hand, waist, etc. They believe that these *bandhanies* are useful at the time of delivery; otherwise the baby will not be born.

HYGIENE AND SANITATION

Individual hygiene and methods of sanitation play a key role in the health status of a society. Although the two villages are thickly populated, the surroundings are not so dirty. But only a few of the houses use *pakha* latrines. Due to the minimum number of doors and windows, houses are devoid of sufficient light and air. The surrounding trees make the houses dark and damp. Most of the families do not possess latrines and they use the nearby jungles or fields for this purpose. The government has provided two tube-wells and four ring wells to the villagers. Some of the families have their own tube-wells. The poorer section of the villagers generally fetch water for household use from their own *kachha* wells. The villagers, in general, do not use any purant to purify the well water.

HERBAL MEDICINE

The villagers have an elaborate knowledge of the medicinal plants and their uses to cure various ailments. This has been continuing in the society from the remote past. Rather than giving an exhaustive list of medicinal plants used by the villagers, let us consider a few of them here and their interesting effects.

1. **Leaves of guava and pineapple** to cure worms of the babies.
2. **Titaphul or Bahka** (*Justicia adhatoda*) leaves for cold and cough; their flowers to cure worms.
3. **Bhedailata** (*Echiles fructescene*) leaves for treating dysentery.
4. **Turmeric rhizome** hot paste for curing sprain, juice for killing worms.

CONCLUSION

The reproductive health condition of the Tiwa women of the two villages, namely, Manipurgaon and Mikirgaon are good. Though a less number of them take vaccines, iron tablets, folic acids during pregnancy, yet the overall health condition of them is good. In this regard, food habit plays an important role. They generally take naturally available vegetables which provide them necessary requirements for body-build. A large number of people still practise their traditional way of delivery. It is because of their wrong concept towards modern procedures of delivery.

Supernatural beliefs play an important role in this regard. For curing of disease, they use locally available herbs. Only in case of major illnesses do they visit the Morigaon Hospital. The economic condition of the people is not that good. Most of them are agriculturists. Perhaps their low income forces them to follow their traditional way of delivery. But their unhygienic living condition, poor sanitation and their cultural beliefs adversely effect their reproductive health.

BIBLIOGRAPHY

Basu, Amitabha. 1990. 'Anthropological Approach to Tribal Health', in Buddhadeb Choudhary (ed.), *Tribal Demography and Development in North-East India*. Delhi: B.R. Publishing.

Choudhary, Buddadeb (ed.). 1986. *Tribal Health: Socio-Cultural Dimensions*. New Delhi: Inter-India Publication.

Medhi, Brinchi K. 1994. 'Health Cuture in a Kaibarta Village'. *Bulletin of the Department of Anthropology*. Vol. VIII, Gandhi University.

Singh and Shukla. 2003. 'Health and Disease among Bhokas: A Tribal Community of Uttranchal State'. *Indian Journal of Physical Anthropology and Human Genetics*. Vol. 22.

12

STIGMA OF HANDIGODU DISEASE: A CASE STUDY FROM KARNATAKA

H.K. Bhat

INTRODUCTION

Stigma is a sign of blemish or taint as upon a slave or a criminal and it is also a mark of infirmity or disease (*Webster's Dictionary*). According to the *Oxford Advanced Learner's Dictionary*, it means "feelings of disapproval that people have about illness or ways of behaviors."

Stigma has to be understood in cultural as well as temporal contexts. Cultural perceptions of the physical or social conditions vary from culture to culture. For example, alcoholism is a stigma in some cultures whereas it is an accepted behaviour in others. Further, within the same culture, stigma may vary from time to time. For example, the divorced status of a man or a woman may be stigmatized in some cultures whereas in others it may be idealized. In this paper, an attempt has been made to describe the origin of Handigodu disease, its epidemiology and the social stigma attached to it.

ORIGIN OF THE DISEASE

Handigodu disease is a peculiar orthopaedic problem suffered mainly by Chennangi and Chaluvadi communities (both Scheduled Castes) in Shimoga and Chikmagalur districts of Karnataka.

The first case of the disease was brought to medical attention at the government hospital, Sager, by Shri H.M. Chandrashekar Bhat of Handigodu village in January 1975. The first patient also belonged to Handigodu village and hence the name 'Handigodu disease' (also referred to as Handigodu syndrome). Despite several studies during the last 32 years by eminent scientists, the disease continues to remain a major medical problem.

GEOGRAPHIC FEATURES OF THE AFFECTED AREA

The disease is reported from only two districts of Karnataka, viz., Shimoga and Chikamagalur. There are about 1000 patients recorded in the various hospitals of the two districts. The disease is reported from the region that falls within 13°-14° N. latitude and 75°-76°. E. longitude. Temperature varies from 20°C to 35°C and the average annual rainfall is from 200 to 300 cm.

The important crops of the region are areca nut, paddy, banana, coffee, black pepper and cardamom.

Idigas, Brahmins, Lingayats, Vokkaligas, Vaishyas and Muslims own most of the agricultural lands. In Chikmagalur district, coffee plantations are owned by people and companies from different parts of the country. The Scheduled Castes generally do not own agricultural land. They work as daily wagers in agricultural lands or coffee plantations. About 54 villages in Shimoga district and 47 villages in Chikmagalur district are defected by the disease.

The region which is affected by the disease is known as Malnad (i.e., hilly land). The villages are situated in the midst of forests and valleys. The forest cover in this region has been steadily declining during the last 50 years or so. Deforestation is a major problem. The villages and hamlets are mostly connected by mud roads with moderate communication facilities.

PREVALENCE IN DIFFERENT COMMUNITIES

About 95% of the affected patients belong to Chennagi and Chaluvadi communities. It has affected only one Brahmin (male) who is still alive. The remaining patients belong to Idiga and other communities (such as Samagara, Uppara, Acgaru and Kumbara).

CLINICAL PRESENTATION OF THE DISEASE

(a) Age — Incidence

Age	Incidence
0-5	nil
6-15	15%
>16-	85%

(b) SEX

Male:	45%
Female:	55%

		Incidence
(c)	Symptoms presented:	
	Pain in back, hips, knees and other joints,	80%
	difficulty in walking and running	90%
(d)	Deformity (flexion hip, lumbar lordosis)	40%
(e)	Short stature (short trunk, hands up to knees)	37.5%
(f)	Malformations	2.5%

A large number of patients are not able to squat or sit cross-legged. Most of the patients can attend to their personal needs like taking bath, eating food, defecation, etc. They can also carry out certain agricultural and domestic works. Some patients use stick while walking. A few patients literally crawl within the four walls of the house.

(g) Inheritance:

Pedigree analysis of 234 patients conducted by the scientists of ICMR during 1985-1989 showed the following features:

- Both males and females are equally affected.
- Earliest age of onset of the disease is 10 years, but usually found between the ages of 20 to 60 years.
- At least one parent is affected in most of the cases.
- In some families the disease has affected three generations; no effect on reproduction and longevity.
- New cases continue to occur in the affected families.

MEDICAL STUDIES ON HANDIGODU DISEASE

1. The Government of Karnataka sent a medical team in 1975 to study the problem. They carried out a clinico-radiological study on 45 patients and diagnosed it as an osteo-articular disease predominantly involving hip joints. But the research team could not ascertain its etiology.
2. At the request of the Govt. of Karnataka, ICMR sent a research team from NIN, Hyderabad, in 1976. The research team ascribed the disease to eating of crabs poisoned with pesticides or insecticides (DOT, Folidol and Endrin). They coined the term 'Endemic Arthritis of Malnad', for this disease.
3. The Government of Karnataka sent another research team from Bangalore Medical College in 1982 to study the outbreak of the disease in Chikmagalur district. The scientists recommended conducting of genetic studies.

4. The Director-General of ICMR visited the area in 1985 and instituted a Multidisciplinary Task Force at four centres, viz., Lucknow, Meerut, Hyderabad and Bangalore to carry out research on the following areas:

(i) Epidemiological and Environmental survey
(ii) Nutritional, Metabolic and Endocrine evaluation
(iii) Delineation of Clinical, Radiological and Anthropometric features
(iv) Genetic studies

The ICMR Task Force Project on Handigodu Disease Phase-I (1985-1989) submitted its report in 1989. This team concluded that Handigodu disease is an inherited developmental defect of bones and identified it as spondylo-epiphyseal dysplasia of late onset, Handigodu type with autosomal dominant inheritance. The findings were published in Skeleted Radiology in 1994 (23: 611-619). The team also recommended further studies. Thus, controversy regarding etiology of the disease continued.

The ICMR Task Force Project on Handigodu disease Phase-II (2001-2005): The report is not yet published.

REHABILITATION PROGRAMMES FOR PATIENTS

All the rehabilitation programmes taken up so far have been conceived and executed by the social worker, H.M. Chandrashekar.

1. The Corporation Bank Economic Development Foundation (CBEDF) sanctioned two projects to the victims of Handigodu disease, viz., providing medical care by organizing health camps and rehabilitating the people affected by the disease by training them on weaving activity at a cost of Rs. 2.20 lacs.

A team of doctors were brought from Mangalore to conduct health camps for the affected people. After about 2 years of treatment, the team reported that since the disease has become chronic, it was not responding to allopathic treatment up to the expected level. Then the team tried homeopathic medicines. The result was not encouraging after the homeopathic treatment. The rehabilitation programme was successful to some extent. H.M. Chandrashekar guided the affected people and their family members to 'establish Handigodu Cotton Handloom Weavers Produce and Marketing Co-operative Society Ltd.', to train the affected people. The Society now has more than 200 members. The CBEDF provided ten handlooms for the purpose. Fourteen patients were identified for the training, of which 10 have completed the training and are using looms.

The Corporation Bank, Sager branch provided a loan of about Rs. 2 lakhs on concessional interest rate, to the society for purchasing the raw materials. The Society has produced bed spreads, pillow cases, towels, etc. worth Rs. 5 lakhs. Marketing the products is a big problem for the Society. The member-workers are paid Rs. 35/- per day as salary.

The handloom work is a kind of physiotherapic treatment for the patients since the loom has to be operated by using the feet and hands.

Since 1990, the patients were getting Rs. 100/- per month as pension for the physically handicapped, from the Ministry of Social Welfare, Govt. of Karnataka. The amount was raised to Rs. 200/- per month since 2000 due to the efforts of Chandrashekar.

Other rehabilitation programmes conceived by Chandrashekar are not yet implemented.

STIGMA ATTACHED TO HANDIGODU DISEASE

The majority of the patients (above 90%) suffering from the disease are Scheduled Castes belonging to two endogamous communities, viz., Chaluvadi and Channagi. They are generally referred to as 'Harijans' in the region. Untouchability is still practised in the rural areas of Shimoga and Chikmagalur districts. Harijans live in their own localities, which are segregated from the residential localities of other communities. Harijans do not enter the houses of other Hindu or non-Hindu communities and vice-versa. But social interaction is very much there as Harijans provide agricultural labour to other land-owning communities.

Stigma in the case of Handigodu disease is two-fold. The affected people as well as the disease are stigmatized. Working among the patients and their families attracts public criticism. People belonging to the 'high castes' generally avoid Handigodu patients and their family members. Harijan families with Handigodu patients are also avoided by other Harijan families as they are afraid of contacting the disease. Stigma of the disease has affected the marital and occupational prospects of many young men and women.

I. Problems of the Social worker

H.M. Chandrashekar, a Hovik Brahmin of Handigodu village, comes from a poor agricultural background. He has completed high school education. He is a man with a mission. Chandrashekar discovered the disease in January 1975, and since then he is relentlessly working to uplift the condition of the patients suffering from Handigodu disease. He is also helping the researchers in conducting research on various dimensions of the disease.

Chandrashekar has visited and surveyed Handigodu patients in more than 1200 villages of Shimoga and Chikmagalur districts. He is an encyclopedia on the disease and the patients. Except for the period from 1985-1989 and 2001-2005, he has worked for the welfare of the patients spending his own money. The ICMR had appointed him as a social worker for their projects during 1985-89 and 2001-05, for which he used to receive honorarium. Chandrashekar's sister also worked as a female social worker for ICMR during the periods mentioned above.

Since Chandrashekar and his sister spend their time visiting the Harijan families and helping them, the people from all other non-Harijan communities used to treat them as eccentric persons. In fact, they were isolated socially. During the initial years, Chandrashekar very rarely attended weddings or other rituals conducted in non-Harijan families. But he never misses any important ceremonies or festivals conducted by Harijans.

Chandrashekar has received several awards for his social service. Karnataka Rajyotsava Award was given to him in 1998. He has invested all the award money in a trust to help the disabled people.

Chandrashekar admitted that of late, people are acknowledging the services rendered by him for the upliftment of the disabled Harijan patients. This shows that the stigma attached to the disease is slowly decreasing, especially among the educated people. The Swamiji of Havik Brahmins honoured him recently in recognition of his social service.

II. Problems of the Affected Families

The affected families are facing several problems, especially marital, occupational and social.

Landlords in the region usually avoid labourers suffering from Handigodu disease and their family members. This has affected the income of the affected families.

Marital negotiations usually take place between families with Handigodu patients. It is difficult for a man or a woman suffering from the disease to get a spouse from a family without a patient suffering from this disease.

Hence, Handigodu disease is associated with social, psychological and economic problems.

ACKNOWLEDGEMENTS

Thanks are due to Shri H.M. Chandrashekhar for his generous help. Thanks are also due to Shri N.M. Ramatrishna and Shri P.N. Venugopal for their assistance in the collection of data.

BIBLIOGRAPHY

Agrawal, S.S. *et al.* 1994. 'Handigodu disease: a radiological study'. *Skeletal Radiology* 23: 611-619.

Bapu Rao, S. and B.S. Narasinga Rao. nd. 'Investigations on the possible role of trace elements/minetals in Handigodu syndrome' (Mineographed).

ICMR. 2002. 'Handigodu Disease: a continued challenge'. New Delhi: ICMR (a pamphlet).

The Hindu. 1977. 'Handigodu Syndrome: symptoms, field observations and scientific investigation'. *The Hindu*, 2-10-1977.

13

A BARRIER TO SOCIAL PARTICIPATION FOR LYMPHATIC FILARIASIS PATIENTS

A. Krishna Kumari, K.T. Harichandrakumar
and K. Krishnamoorthy

1. INTRODUCTION

Lymphatic filariasis (LF), more commonly known as elephantiasis, is a devastating parasitic infection in humans, spread by mosquitoes. It remains an important public health problem in India (Das *et al.*, 2000; Das *et al.*, 2001). About 45% of the Indian population live in known endemic areas (WHO, 2000). This tropical disease causes long-term suffering and morbidity as well as high social and economic burden to individuals, communities and the nation (Ramaiah *et al.*, 1997; Gyapong *et al.*, 1996; Ramaiah *et al.*, 2000), impairs mobility, day to day domestic activities (Evans *et al.*, 1993) and sexual and marital life (Dreyer *et al.*, 1997). It also imposes burden on the health care infrastructure in endemic areas (Haddix and Kestler, 2000). Even though the disease does not kill, it is ranked as the second leading cause of disability (WHO, 1995).

Acquired early in life, LF is transmitted between humans by the mosquito during biting. Millions of microscopic microfilariae circulate in the blood and are produced by adult worms, which live for five/six years in the lymphatic system. The worst symptoms of the chronic disease generally appear in adults, and in men more often than in women. In endemic communities, some 10-50% of men suffer from genital damage causing hydrocele (fluid-filled, balloon-like enlargement of the sacs around the testes) and elephantiasis of the penis and scrotum. Lympheodema of legs and arms are common among men and women. Elephantoid vulva and breast swelling up to several times normal size is also reported among women (WHO, 2000).

LF imposes a significant financial burden in endemic areas (Krishnamoorthy, 1999; Ramaiah *et al.*, 2000), as the chronic disabilities affect the individuals in their most productive years. LF patients often find it difficult to find a job, have to forgo many education to the community, can lessen the social handicaps and distress caused by lymphatic filariasis.

The strategy of the Global Programme to Eliminate Lymphatic Filariasis (GPELF) has two components: firstly, to stop the spread of infection (i.e., interrupt transmission), and secondly, to alleviate the suffering of affected individuals (i.e., morbidity control).

The GPELF strategy for disability prevention views disability from two angles, from the medical angle as well as the social angle based on the WHO International Classification of Functioning, Disability, and Health (ICF)'s description of health. The medical model views disability as an individual problem directly caused by a disease requiring medical care of the individual by health professionals. Conversely, the social model views disability as a matter of an individual's full integration into society and prescribes social measures that will create the conditions necessary to enable disabled persons to fully participate in all areas of social life (www.mectizin.org).

The GPELF considers advocacy as an important component to ensure programme implementation. Information on the burden due to LF is crucial for advocacy. This communication reports our observations on the impact of the different health states of filariasis on the social participation of the patients.

2. MATERIAL AND METHODS

2.1. Study Area

The study was carried out in Pondicherry urban agglomeration and two LF endemic villages in Villupuram district of Tamil Nadu state in South India. There were 212 patients detected through physical examination by the health worker through house-visits as well as those who reported at the Filaria Night Clinic at Pondicherry. A total of 174 patients between 15 to 60 years of age were involved in the study. The patients who were working lost their job, or have to switch over to less strenuous works earning less wages because of their disability.

Lymphatic filariasis also exerts a heavy social burden that is especially severe because of the specific attributes of the disease, particularly since chronic complications are often hidden and are considered shameful. For men, genital damage is a severe handicap leading to physical limitations and social stigmatization. For women, shame and taboos are also associated with the disease. When affected by lymphoedema, they are considered undesirable and when their lower limbs and genital parts are enlarged they are severely stigmatized; marriage, in many situations an essential source of security, is often impossible. Clinical signs of LF result not only in physical but also in psychological damage (WHO, 2000).

The socio-economic component of the disease is perhaps the most neglected aspect of lymphatic filariasis research. Even though one-third of the world's population lives at risk in India (WHO, 1984; Anonymous, 1989), filariasis has not received its due attention because the socio-economic impact is underestimated and largely unrecognized. Also, information on the social and psychological problems it causes is scanty (Suma and Shenoy, 2003). Scientific study and documentation of the social impact of diseases will not only enable one to understand the extent of the problem but also help to devise means to overcome them (YCRC, 1990).

With dramatic advancement in research towards the development of tools and strategies, this disease has now been rated as eradicable. Accordingly the Global Programme for the Elimination of Lymphatic Filariasis (GPELF) was launched in 2000, and proposed to cover 350 million people in 46 countries in 2005 (Ottesen *et al.*, 2004). It has been classified into different health states (clinical manifestations) following WHO classification criteria (WHO, 1992).

2.2. Health States

Lymphatic filariasis is characterized by a wide spectrum of clinical manifestations. The clinical course of filariasis can be divided into asymptomatic, acute and chronic stages, generally progressing in that order. The asymptomatic stage is characterized by the presence of microfilariae in the blood, although there are no clinical manifestations of filariasis. The acute stage, Adenolymphangitis (ADL), is characterized by recurrent attacks of fever associated with inflammation of the lymph nodes and lymph vessels. The acute course of filariasis may last for several days or up to 4-6 weeks and usually subsides spontaneously following bed-rest. Lymphoedema is quite often present in these episodes, but subsides after each episodic attack (WHO, 1987).

Based on the criteria recommended by the World Health Organization (WHO, 1992), six chronic manifestations of lymphatic filariasis, namely, lymphoedema grade 1 (L1 - pitting oedema of the limb that is reversible on elevating the limb), lymphoedema grade 2 (L2 - pitting/non-pitting oedema that is not reversible on elevation of the limb and the skin is normal), lymphoedema grade 3 (L3 - non-pitting oedema of the limb, not reversible on elevation and the skin is thickened), lymphoedema grade 4 (L4 - Non-pitting oedema with fibrotic and verrucous skin changes), hydrocele grade I(Hl - <15 cm diameter) and hydrocele grade 2 (H2 - ≥15 cm diameter) and ADL associated with chronic manifestations were considered as health states of LF.

2.3. Instrument used for Measuring HS

A 7 Domain 5 Level (7D5L) descriptive system was used to describe the different health states of LF. The 7D5L instrument is designed to assess the outcomes of the disease on physical, mental and social health. The social domain is assessed in terms of maintaining relationships, visiting/inviting relatives and friends, attending or hosting social functions, etc. The severity levels of the social domain ranging from 'no problem' to 'very severe problem' were predefined and coded as: 1- no problem, 2- mild problem, 3-moderate problem, 4- severe problem and 5- very severe problem. The effect on social participation was assessed based on questions whether the patient was able to maintain social relationships, visit/invite relatives and friends, attend/host social functions, etc. The degree of severity was assessed with a five-point scale that ranged from no difficulty to profound difficulty experienced by the patient in each of the domains. In the domain of social participation, mild severity means individuals involve in social activities, but with hesitation. Moderate severity means that social participation is restricted. A rating of 'severe' indicates maximum avoidance of social activities and 'extreme' severity means total avoidance of social participation. Based on the above domain and severity levels, a matrix was prepared and used for recording the response from the patients.

The interview was conducted in the patient's domestic settings after briefing them about the purpose of the interview. Informed written consent was obtained from each respondent before the interview. Confidentiality was maintained for the data obtained as per ethical guidelines (ICMR, 2000). Each patient was asked to state the level of severity in social participation. The severity level of each domain was transformed into score by giving values for no problem as 0 and for mild, moderate, severe and extreme severe problem as 1, 2, 3 & 4 respectively. The values were computerized for further analysis. The mean severity score for social participation was calculated for each health states.

3. RESULTS

A total of 154 patients with chronic manifestation of different stages of oedema (L1 n = 26, L2 n = 31, L3 n = 23, L4 n = 20, H1 n = 27, H2 n = 23) and

24 cases with acute episodes of ADL were interviewed. There were 84 women and 90 men and their mean age was 45.8 and 40.8 years respectively. Most had no formal education and were agricultural labourers (n = 55), farmers (n = 31), weavers and artisans (n = 17), government employees (n = 8) or unemployed (n = 63).

Acute episodes were the cause for physical incapacitation, limiting the activities such as mobility, usual activities and self care. On an average, the patients with acute attack of ADL suffer for three days during when they would be completely bed-ridden and in a state of delirium. Their social participation is affected mainly due to physical incapacitation.

Patients with initial stages of chronic manifestation (H1 and L1) of LF reported mild effects on social participation and hesitated to take part in social activities. The main reason attributed was that travelling and certain food items served during feasts would aggravate pain and swelling. This could result in ADL, a co-morbidity of these patients. These patients reported mild problems with mobility and usual activity.

In the case of chronic manifestations, in addition to incapacitation, their social participation was very much affected and the severity levels showed an increasing trend with increasing grades of oedema and deformity. Acute attacks of ADL associated with chronic manifestations were treated as co-morbidity and considered while assigning severity levels by the respondents.

There was total avoidance of social participation by the patients with grade 4 oedema. This was mainly because of stigma and shame associated with the gross disfiguration. One patient said, "I won't even go to the nearby junction because I feel ashamed." Another patient said, "I am physically and mentally broken due to this disease and totally avoid all social activities". Patients in the health state of L3 restricted their social participation. One respondent said, "I attend only death ceremonies. There no one will tease me." Patients in the health state of L2, took part in social activities but with hesitation. Many ladies said they had to wear their saris lower than usual to hide their enlarged limbs. An 18-year-old boy said he was hesitant to participate in any function because he could not wear the type of dress of his choice. He could wear only *dhoti* that would hide his enlarged legs. He said he would feel irritated and dejected when his friends tease and make fun of him for wearing *dhoti*.

4. DISCUSSION

Lymphatic filariasis imposes a heavy morbidity burden in terms of mobility, pain, usual activity and self care. The present study shows that the disease also restricts the social participation of affected individuals. The severity increases with the advancement of the disease. ADL, an acute manifestation of LF, is self-limiting and temporary. It imposes physical incapacitation for an average of three days per attack and its impact is mainly due to pain and discomfort, making them incapable to take part in social activities. People remain bed-ridden and are compelled to seek medical care (Nanda and Krishnamoorhty, 2003). However, it is a precipitating factor (Pani *et al.*, 1995) in the natural history of LF and progression of disease. Their social participation is affected mainly due to physical incapacitation.

Chronic manifestations limit social activities mainly due to stigma. In addition to difficulties faced in physical domains such as mobility, usual activity and self care, pain and discomfort as well as psychological effect, stigma affects the patients in performing social activities. Higher the grades of lymphoedema and hydrocele, greater is the interference reported with the social life of the individual. Patients with large hydroceles or elephantoid legs with massive skin folds and dreadful odour were victims of teasing and social stigmatization. Individuals with gross deformity were dejected, and they avoided all social activities and led an isolated life within the family. Individuals with lower grades of lymphoedema restricted their social activities to essential and unavoidable contacts, for fear of inadvertently eating foods that would induce ADL attacks, concern about ADL attacks after long hours of travel, and shame of exposure the deformity. Sons and daughters of LF patients had difficulty in getting suitable marriage proposals. It was also noticed that house-owners were reluctant to rent out their houses to LF patients.

Limb care and surgical corrections of hydrocele are the two major methods recommended by WHO (2006) for morbidity management. While the former is focused on the continuous practice of leg washing and maintaining good personal hygiene, the latter offers permanent solution. Though the current package on limb care emphasizes this procedure, there is no scope for counselling. The present study clearly shows that counselling should be a part of the training of patients in limb care as it affects the social life of individuals. Further, the results of the study can be used in the messages designed for advocacy while implementing programmes to eliminate lymphatic filariasis.

Medical test for physical fitness is part of the recruitment procedures in certain services, especially the defence. Patients with hydrocele are considered unqualified and hence this causes a great concern for unemployed youths. This, again, is a social discrimination resulting from stigma. The message of surgical correction for permanent cure of hydrocele patients should be included in the social mobilization package for morbidity

Table No. 1
Number of Patient-respondents Participated in the Assessment of Health Status of Lymphatic Filariasis

Health State	Number of Patient-respondents	Median Age (Range) in Years
Adenolymphangitis (ADL)	24	43 (18-57)
Lymphoedema grade 1 (L1)	26	45 (18-60)
Lymphoedema grade 2 (L2)	31	50 (20-60)
Lymphoedema grade 3 (L3)	23	42 (22-60)
Lymphoedema grade 4 (L4)	20	45 (20-56)
Hydrocele grade 1 (H1)	27	37 (15-59)
Hydrocele grade 2 (H2)	23	43 (33-60)

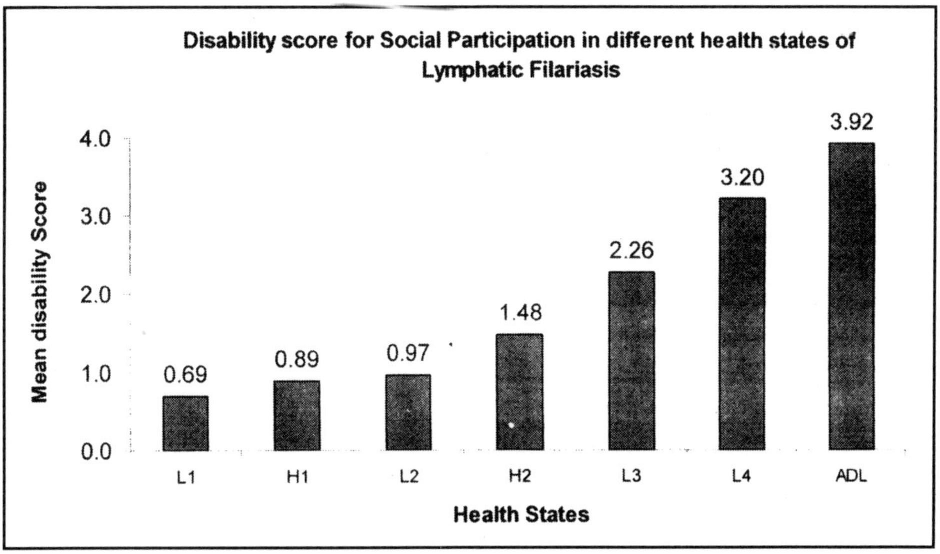

Fig. 1. Disability Score for Social Participation in Different Health States of Lymphatic Filariasis.

management under the GPELF, so as to relieve these patients from the fear of disqualification. Programmes to eliminate lymphatic filariasis should consider these findings while developing implementation plans.

Rehabilitation and counselling are required for diseases which require long-term treatment. In the case of lymphatic filariasis, prevalence of advanced cases with grade 4 is very low (Pani and Srividya, 1995; Pani *et al.*, 1990). Long-term chemotherapy is not required for these patients. They can be given counselling by health workers during the house visits. While counselling can improve the mental health of the affected individuals, their social health can be improved through educating the community on filariasis. The root cause of the stigma and social isolation of LF patient is the ignorance of the community about the disease. Patients, their families and the community are generally unaware of the cause of the disease and control measures. In endemic areas, the community should be taught how to reduce infections and also to help those who are already suffering due to the disease (Davies, 1975). Health education in endemic areas will also help to increase drug consumption during MDA and achieve vector control through environmental management. Such an approach will lend to achieve the global goal of elimination of LF by 2020 (WHA, 1997).

BIBLIOGRAPHY

Anonymous. 'Filariasis in India'. Editorials in *National Med. J. India* 3 (1989) 1.

Das, P.K., K.D. Ramaiah, D.J. Augustian and Ashok Kumar. 2001. 'Towards elimination of lymphatic filariasis in India'. *Trends in parasitology*. Vol. 17, No. 10, pp. 457-460.

David, G. Addiss and Charles Mackenzie. 2004. 'LF DISEASE-CLINICAL MANAGEMENT'. *Am. J. Trop. Med. Hyg.* 71 (5 suppl.), pp. 12-15: http://www.ajtmh. orgicgi/content/full/71/5_suppl/12.

Davies, J.B.M. 1975. 'Community Health, Preventive Medicine and Social Service'. *Bailliere Tindall*. London, p. 129.

Dreyer, G., J. Noroes and D. Addis. 1997. 'The silent burden of sexual disability associated with lymphatic filariasis'. *Acta. Trop.* 63: 57-60.

Evans, D.B., H. Gelband and C. Vlassof. 1993. 'Social and economic factors and the control of Lymphatic filariasis: a review'. *Acta. Trop.* 53: 1-26.

Gyapong, J., M. Gyapong, D.B. Evans, M.K. Aikins and S. Adjei. 1996. 'The Economic Burden of Lymphatic Filariasis in northern Ghana'. *Ann. Trop. Med. Parasitol* 90, 39-48.

Haddix, A.C. and A. Kestler. 2000. 'Lymphatic filariasis: economic aspects of the disease and programmes for its elimination'. *Trans. R. Soc. Trop. Med. Hyg.* 94: 592-93.

Indian Council of Medical Research. 2000. 'Statement of general principles on ethical considerations involving human subjects', *Ethical Guidelines for Biomedical Research on Human Subjects*, pp. 1-8.

Krishnamoorthy, K. 1999. 'Estimated cost of acute adenolymphangitis to patients with chronic manifestations of bancroftian filariasis in India'. *Ind. J. Pub. Health* 43: 58-63.

Nanda, B. and K. Krishnamoorthy. 2003. 'Treatment seeking behaviour and costs due to acute and chronic forms of lymphatic filariasis in urban areas in south India'. *Trop Med Int Health* 8(1): 56-9.

Ottesen, E.A., G.J. Weil, P.J. Lammie *et al.* 2004. 'Towards a strategic plan for research to support the global program to eliminate lymphatic filariasis. Summary of Immediate needs and opportunities for research on lymphatic filariasis'. *Am. J. Trop. Med. Hyg.* 71(5 Suppl.): 3-46. http://www.mectizan.org/newsletters/mpn36eng.pdf. Mectizin programme notes Issue 36. Review of Progress in Disability Prevention Activities Associated with Lymphatic Filariasis in Africa (Accessed on 5.2.07).

Pani, S.P. and A. Srividya. 1995. 'Clinical manifestations of bancroftian filariasis with special reference to lymphoedema grading'. *Indian Journal of Medical Research* 102: 114-118.

Pani, S.P., K. Krishnamoorthy, A.S. Rao and J. Prathiba. 1990. 'Clinical manifestations in Malayan filarial infection'. *Indian Journal of Medical Research* 91: 200-207.

Ramaiah, K.D., K.N. Vijay Kumar, K. Ramu, S.P. Pani and P.K. Das. 1997. 'Functional impairment caused by lymphatic filariasis in rural areas of south India'. *Trop Medicine & International Health*. 2: 832-838.

Ramaiah, K.D., P.K. Das, E. Michael and H. Guyatt. 2000. 'The Economic Burden of Lymphatic Filariasis in India'. *Parasitology Today* 16: 251-253.

Suma, T.K. and R.K. Shenoy. 2003. 'A qualitative study of the perceptions, practices and socio-psychologial suffering related to chronic brugian filariasis in Kerala, Southern India'. *Annals of Tropical Medicine and Parasitology* 97: 839-845.

Vector Control Research Centre. 1990. 'Misc Publ 16 on Proceedings of a seminar on future research needs in lymphatic filariasis'. Pondicherry: VCRC.

World Health Assembly. 1997. 'Elimination of Lymphatic Filariasis as a public health problem'. *World Health Assembly*, Geneva A50.291997.

World Health Organization. 1984. 'Fourth report of WHO expert committee on Lymphatic filariasis'. Filariasis. WHO Tech Rep. Ser. 702. P 112.

——. 1992. 'Control of lymphatic filariasis. A manual for health personnel', Geneva: World Health Organisation.

——. 1992. 'Lymphatic filariasis. The disease and its control. Fifth report of the WHO Expert Committee on Filariasis'. WHO Technical Report Series 821, 1-71.

——. 1995. 'The state of world health. World Health Report- Bridging the gaps' (WHO, Geneva): 1-42.

——. 2000. 'Eliminate Filariasis: Attack Poverty'. The Global Alliance Lymphatic Filariasis. Proceedings of the First Meeting. WHO/CDS/CPE/CEE/2000.5.

——. 2000. 'Lymphatic filariasis'. Fact sheet. Number 102. http://www.who.int / mediacentre/factsheets/fs I 02/en/index.html. (Accessed on 05.02.07).

——. 2006. 'Global Programme to Eliminate Lymphatic Filariasis'. *Wkly Epidemiol Rec.* 2006, 81 (22): 221-232.

14

HUMAN BEHAVIOUR AND PERSISTENCE OF MALARIA

L.K. Das and S.P. Pani

1. INTRODUCTION

1.1.

Malaria, a disease of antiquity, has proved to be a formidable deterrent to the cultural and socio-economic progress of man in the tropical, sub-tropical and monsoon zones of the world. History is replete with instances of devastation caused by this disease. Wide distribution and its intensity of transmission in India were important factors for slow economic, scientific and industrial progress in the country during the last two centuries. However, when the biological characteristics of vector-borne diseases began to be unravelled through pioneering research initiated by Sir Ronald Ross, the mechanism of transmission of human malaria began to be revealed. Malaria is one of the major public health problems in the developing countries. It has affected 88 countries and 3 billion people are at risk. This disease is a devastating scourge in the lives and livelihoods of billions living in tropics and sub-tropics. A recent estimate suggests that about 500 million to 5 billion clinical episodes and over 3 million deaths occur annually.

During 1996 also, India contributed 83% of total malaria cases in SE region. Thus, around 80% of reported cases in the region are being contributed by India.

1.2. Situation in India

Malaria has been known in India from times immemorial. J.A. Sinton, the first Director of the Malaria Institute of India, estimated in 1935 that at least 100 million people suffered from malaria in India and about one million

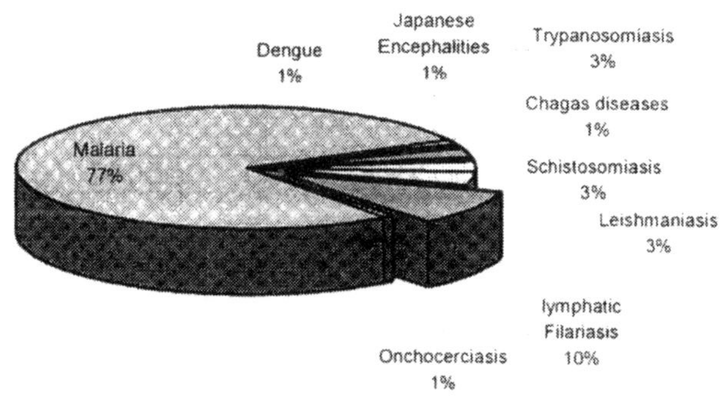

Fig. 1. Contribution (%) of each Vector Borne Disease (VBD) to the global burden due to VBD-Global 1- Malaria; 2- Onchocerciasis; 3- Lymphatic Filariasis; 4-Leishmaniasis; 5- Schistosomiasis; 6- Chaga's diseases; 7- Trypanosomiasis; 8-Japanese Encephalitis; 9-Dengue

deaths occurred annually in endemic areas. In 1947, after independence, it was estimated that 75 million people suffered from the disease every year. The mortality, as direct result of disease, was placed at 0.8 million per annum. By 1965 the annual malaria incidence fell from 75 million cases to an all time low of 0.1 million cases. No deaths were recorded. There were setbacks from 1968 onwards and highest ever post-eradication incidence of 6.47 million cases was recorded in 1976. Presently, the incidence is stabilized at approximately 2 million cases annually.

2. MATERIAL AND METHODS

2.1. Study Area and population

The erstwhile Koraput was the largest district (17° 50′ and 20° 30′N and 81° 27′ and 84° 10′E), in Orissa state, with an area of about 27, 900 sq. km and a population of about 3 million. The population density is relatively low (111/sq. km). About 60% of the total population was of tribal origin. Out of 62 Scheduled Tribes in the state of Orissa, as many as 51 were found in this district. Gadavas, Kondhs and Koyas were selected for the study, as they were numerically large in the district. Their population is the maximum in the eastern part of Koraput district. They live mostly in the north-eastern region of the district. Out of total Kondh population in Orissa state, about 35% live in this district. The Koyas are a branch of Gondhi-speaking people distributed in the southern part of the district, i.e., the present Malkangiri district and the neighbouring state, Andhra Pradesh. The houses in villages

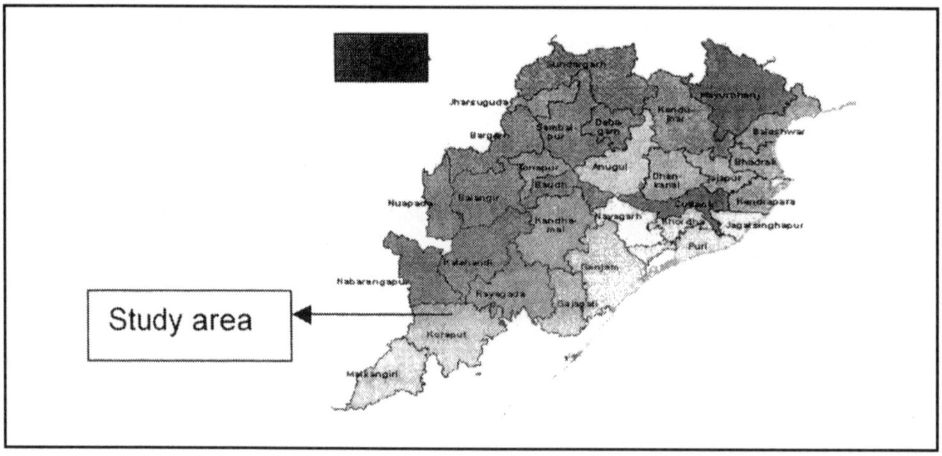

Picture 1. The State of Orissa with Study Area

are with thatched roofing and mud-plastered walls, and are usually in-groups to form a village. Apart from the main cluster of houses, some areas have a few scattered holdings close by (hamlets). The majority of the population is of tribal background, who are economically poor. The menfolk go for cultivation of mainly paddy and women go for collection of forest products, usually *sal* leaves and firewoods. The health care facilities are poor and the first 'medical' assistance is from quacks and *disharis* (traditional healers) (Rajagopalan and Das, 1990).

2.2. Malaria Situation in the Study Area

The villages in the district are highly endemic for malaria. The annual parasite incidence ranged from 25 and 45 during the last five years. More than 93% of the malaria cases are caused by *Plasmodium falciparum.* Malaria

Picture 2. A Typical Tribal Village

Picture 3. Typical Tribal People

transmission occurs throughout the year. The incidence is low during summer and starts increasing during rainy months and peaks in Nov.-Dec. The topography and dynamics of malaria transmission in the area have already been described (Rajagopalan *et al.*, 1990). Villages situated on hilltops and at the foot of hills are hyper-endemic for malaria and experience perennial transmission. Villages situated on plain lands and riverbanks are hypo-endemic for malaria and experience seasonal transmission in rainy season (Rajagopalan *et al.*, 1990; Jambulingam *et al.*, 1991).

2.3. Survey techniques

A census was carried out in all the selected villages to identify the important demographic characteristics of the study population. The general beliefs and faiths of the people with respect to health and disease were studied by questioning a cross-section of the people (mostly, observational and informal discussion) by a team of workers. The frequency of mud plastering was observed on 3, 7, 15, 21 and 30 days after spray.

Qualitative data for traditional beliefs and practices for personal protection measures against mosquito bites were collected using key informant and in-depth interviews. They were supplemented by focus group discussions. Trained field workers of the same ethnic group were employed to assist in conducting the interviews in local tribal language. Most of the interviews were tape-recorded. To obtain quantitative data, a structured questionnaire was developed, based on the qualitative information. A minimum of 20% of the population in the study villages was interviewed following an age (>15 years of age) and sex stratified proportionate sampling protocol. Direct observation of certain behavioural patterns and traditional method of protection from mosquito bite was also made.

2.4. Interviews

A total of 18 key informant interviews, 71 in-depth interviews and 15 focus group discussions were conducted in the study villages. The quantitative survey was conducted among 455 persons.

2.5. Data Analysis

Typed verbatim transcripts were obtained from audiotapes of the group discussions and in-depth interviews and the data were analysed using the Text Base Alpha Programme. The quantitative data were analysed using EPI-INFO software. Differences between proportions found in different groups were tested for statistical significance using the X^2 test.

3. RESULTS

3.1. Causes of Persistence of Malaria

Studies carried out by VCRC during last 20 years have revealed that the causes of persistence of malaria in the hard-core areas can be broadly classified into the following categories:

(1) Technical aspects (ecological conditions of the area, problems related to malaria parasite, problems related to mosquito vectors).
(2) Human behavioural aspects.
(3) Administrative, operational and logistical aspects in relation to the existing control programmes in including health care facilities.

3.2. Human Behaviour and Persistence of Malaria

Social Customs and Persistence of Malaria

Faiths and beliefs

The traditional beliefs and faiths relating to illness have a deep cultural root, particularly among the tribals. Traditionally, there are three possible causes of illness. They are: a) vengeance of divine beings (Goddesses or 'Thakurani' as known locally) due to the wrong doings of individual, b) difficult times related to the planetary positions, and c) illness as a result of witchcraft or black magic. These beliefs are not only prevalent among illiterate tribals but also among the literates. The probable cause (of the three mentioned above) of illness is decided from the symptomatology of the disease or according to the strongest belief of the person. For example, any disease with fever and skin rashes (like chicken pox or measles) is taken to be definite sign of the vengeance of goddess.

Treatment seeking behaviour

An average 3.5 days is the time lag in seeking treatment in these areas resulting in high morbidity and mortality. This is due to lack of trained personnel, distance from health centres, lack of proper transport and non-availability of proper medicines. They believe in three different modes of 'treatment' according to the three possible causes of disease and there are, therefore, three types of traditional healers. The help of a *Pujari* or priest is sought if the illness is believed to be due to divine vengeance; a *Gunari* or astrologer is approached if bad planetary positions are thought to be responsible; and a *Dishari* (one who is believed to drive away evil spirits) is called when witchcraft or black magic is felt to be the source of illness. The tribals approach the nearest hospital or health care personnel only when al the above methods fail, that too when they can afford to transport the sick.

Of the three traditional healers, the *Disharis* appear to have developed some system of their own for the diagnosis and treatment of the sick. They have traditional scripts written of palmyra leaves, which deal with symptomatology and appropriate recipes. They recognize altogether 11 types of fevers. Of these, three appear to be related to malaria (Pani, 1990).

(i) Pali or Kampa Jwara
(ii) Bata Salisama Jwara
(iii) Sannipata Jwara

Mud plastering of sprayed surfaces

The tribals living in mud wall houses traditionally plaster the walls with coloured mud decoratively. This practice probably evolved from the need to prevent the appearance of cracks and crevices in the walls. Study of normal frequency of mud plastering in two villages showed that floors are plastered almost daily, indoor walls almost every week, and outside walls and cattle sheds on ceremonial occasions (Tables 1 & 2).

Non-acceptance of indoor residual insecticide spray

Reasons for refusal of residual insecticide spray: In about 33% of the holdings there was refusal. The following were the important reasons for the refusals: religious causes (43%, since most tribals do not permit any outsiders to

Table No. 1
Frequency of Mud Plastering of Insecticide Sprayed Surfaces in the Study Area

Nature of Surface	Frequency of Mud Plastering (% of total holdings)						
	Daily	**Weekly Twice**	**Weekly Once**	**Fort-nightly**	**Monthly**	**Once in 3 Months**	**Cere-monial**
Floor	64.3	31.5	4.2	0.0	0.0	0.0	0.0
Indoor wall inner portion	0.0	76.2	14.7	3.5	1.4	0.7	3.5
Indoor wall upper portion	0.0	25.9	42.0	4.2	2.8	1.4	23.8
Outer wall	0.0	4.2	12.6	8.4	1.4	0.7	72.7
Cattle shed	0.0	0.0	1.3	0.0	0.0	11.8	85.5

Table No. 2
Proportion of Human Dwellings/Cattle Sheds, Mud-Plastered on Different Days following Residual Insecticide Spraying

Days of Follow-up	Number Surveyed	No. Completely Plastered (%)	No. Partially plastered (%)	Total Plastered (%)
Human dwelling				
3	114	53 (46.5)	30 (26.3)	83 (72.8)
7	126	75 (59.5)	31 (34.6)	106 (84.1)
15	127	90 (80.9)	27 (21.3)	117 (92.1)
21	127	103 (81.1)	20 (15.7)	123 (96.8)
30	130	128 (98.5)	2 (1.5)	130 (100.0)
Cattle sheds				
3	71	1 (1.4)	1 (1.4)	2 (2.8)
7	71	16 (22.5)	1 (1.4)	17 (23.9)
15	71	23 (32.4)	1 (1.4)	24 (33.8)
21	71	25 (35.2)	2 (2.8)	27 (38.0)
30	71	61 (85.9)	1 (1.4)	62 (87.3)

enter into their *Pooja* or prayer rooms), apprehensions of bedbug nuisance following spray (24%), belief that spray has no use (9%), bad smell (6%), presence of sick in the family (6%) and fear of spoiling stored foodgrains (4%).

Sleeping habits: Outdoor sleeping habits

The sleeping habits of villagers depend on the season. In summer, the majority of Koyas (99%), 59% Gadavas and 50% Kondhs sleep in verandahs and outdoors, while others, especially females, sleep indoors. In rainy and winter seasons, the majority sleep indoors (Gadavas 87-91%; Kondhs 70-80%; Koyas 70-76%). About 28% of Kondhs, 6% Gadavas and 2% Koyas stay in the fields during night for guarding their crops. Among Koyas, nearly 6% sleep outdoors even during winter. In colder months, the residents sleep near fire for warmth and protection from cold.

Housing pattern

The majority of the houses (>90%) are made up of mud wall and thatched roof (bamboo and straw). A few are with cemented brick walls and tiled roofs. Remaining is houses with mud walls and roofed with tin sheets. The houses generally have two rooms, one room with false ceiling for conducting *pooja*, storing grains and other household materials and the other is a kitchen, which is also used for sleeping during cold and rainy season. Invariably, every house has a verandah, which is the most frequently used place by the inhabitants including for sleeping during hotter months. All the houses have one door system. There are no windows in the huts and there is a gap of 15-30 cm between the wall and roof (eaves) for ventilation and light.

Personal protection measure against mosquito bites. Traditional practice against mosquito biting

The majority of the people did not adopt any specific protection measure except simply swatting the mosquitoes that land on the body. The next commonly used method was covering the body with cloth while sleeping. Other methods included fumigation of smoke in the house by dry leaves and twigs of the local herb of the genus *Ocimum,* application of oil, locally called 'Kusum' *(Carthamus tinctorins),* 'Mohulo' *(Maduca longifolia)* and 'Koranj' *(Pongamia glabra)* on the body while working in the field or forest. In summer, majority (Gadavas: 80.8%, Kondhs: 78.3% & Koyas: 84.6%) did not use any protection measure. In rainy and cold months, 60-70% of the people used any one of the measures or sometimes in combination.

During daytime when visiting forest, significantly larger number of Kondhs (33%) used oils as repellents compared to Gadavas (7%) and Koyas (13%) (x^2 = 34.4; $P<0.01$). Others stated that they would wear clothes or physically drive away the mosquitoes. When sleeping in the field for guarding the crop, the use of smoke was more common than applying oils.

Use of other protection measures: Mosquito nets

About 2% of Gadavas, 52% of Kondhs and 13% of Koyas reported that they had never seen a mosquito net. A few told that they had seen mosquito nets in weekly markets/nearby town. The Koyas had seen mosquito nets in the neighbouring villages, where Bengali settlers had been traditionally using mosquito nets (Das *et al.,* 1993). The majority felt that they could not afford to buy mosquito nets.

The low level of awareness about mosquitoes as a malaria vector among the three major tribes in the district is one of the reasons for the persistence of malaria in the area. Therefore, intense IEG Programme oriented towards building up awareness about mosquitoes and their role in malaria transmission is highly essential before introduction of personal protection measures.

4. DISCUSSION

It is difficult for governments, particularly in developing countries, to provide health services for all their people. Recognising that, any preventive measures, promotive and curative procedures do not require extensive professional training. The concept of community participation, where individuals and families assume responsibilities for their own health, are being encouraged. Accordingly, local people are required to cooperate with the professional health workers in dealing with the health problems prevalent in the community. Socio-cultural practices and responses are critical in achieving this goal. The persistence of malaria in several parts of the world is closely associated with socio-cultural practices and human behavioural aspects (Reuben, 1989). These include 'folk concept of malaria', 'folk remedies of malaria', 'role of traditional practitioners' and 'migration and economic activities' (WHO, 1988). There are several reasons for resorting to treatment from the traditional healers, the most important being illiteracy and ignorance followed by poverty. These treatments are thought to be a part of the tradition followed for generations. The community accepts the healers because he belongs to the same community and charges only in kind, which the patients can afford.

The persistence of malaria in the republic of Somalia was due to migration of people related to traditional pastoralism (Prothero, 1984) and in Rameswaram Islands in South India due to seasonal movement of population for fishing (Rajagopalan *et al.*, 1984; Jambulingam *et al.*, 1986). It is well known that malaria is a local problem. Therefore, control strategies based on generalization are likely to be followed by most disastrous consequences. As such, strategies for control of malaria in India for hard-core areas like Orissa, Chhatisgarh and north-eastern states have to be developed for successful control of malaria.

BIBLIOGRAPHY

Das, P.K., L.K. Das, S.K. Parida, K.P. Patra and P. Jambulingam. 1993. 'Lambdacyhalothrin treated bed nets as an alternative method of malaria control in tribal villages of Koraput District, Orissa State, India'. *South East Asian Journal of Tropical Medicine and Public Health* 24: 513-521.

Jambulingam, P., S.S.S. Mohapatra, P. Goverdhini, Lalit Kumar Das, A. Manoharan, S.P. Pani and P.K. Das. 1991. 'Micro level epidemiological variations in malaria and its imptications on control·strategy'. *Indian Journal of Medical Research* 93: 371-378.

Jambulingam, P., S. Sabesan, K. Krishnamoorthy, G. Rajendran, N. Pradeep Kumar and P.K. Rajagopalan. 1986. 'Malaria Control in Rameswaram Island in South India'. *ICMR Bulletin* 16(8): 89-94.

Pani, S.P. 1990. 'Epidemiology of malaria persistence in Koraput district, Orissa state'. *M.D. Thesis in Epidemiology,* Pondicherry: Pondicherry University.

Prothero, R.M. 1984. 'Malaria: Socio-economic and geographical problems'. *Proceedings of the Indo-UK workshop on Malaria.* New Delhi, November 14-19, 1983, Malaria Research Centre (ICMR), pp. 41-52.

Rajagopalan, P.K. and P.K. Das. 1990. 'The problems of malaria control in tribal areas'. *ICMR Bulletin* 20: 41-46.

Rajagopalan, P.K., P.K. Das and S.P. Pani *et al.* 1990. 'Parasitological aspects of malaria persistence in Koraput district Orissa, India'. *Indian Journal of Medical Research* 91: 44-51.

Rajagopalan, P.K., P. Jambulingam, S. Sabesan, K. Krishnamoorthy, S. Rajendran, K. Gunasekaran and N. Pradeep Kumar. 1984. 'Population Movement and Malaria Persistence in Rameswaram Island'. *ICMR Bulletin* 14(11): 125-33.

Reuben, R. 1989. 'Obstacles to malaria control in India-the human factor', in M.W. Service (ed.), *Demography and vector borne diseases.* Florida: CRC Press, pp. 144-154.

World Health Organization. 1988. 'The impact of folk healers on the performance of malaria volunteers in Thailand'. *Socio and Economic Research Project Reports* 4: 51-53.

15

A FEELING OF FULFILMENT AT MENOPAUSE: A STUDY AMONG URBAN MENOPAUSAL WOMEN IN MAHARASHTRA

Aarti Kaulagekar Nagarkar

INTRODUCTION

A total of 130 million Indian women are expected to live beyond the menopause into old age by 2015. The menopause is emerging as an issue owing to rapid globalization, urbanization, awareness and increased longevity in urban middle-aged Indian women, who are evolving as a homogeneous group. Improved economic conditions and education may cause the attitude of rural working women to be more positive towards menopause. However, most remain oblivious of the short- and long-term implications of the morbid conditions associated with middle and old age, simply because of lack of awareness and unavailability or ever-increasing cost of medical and social support systems. Evidence-based medicine is accessible to still only a few Indian women. Most menopausal women go untreated or use unproven alternative therapies (Sengupta, 2003).

Anthropological and cross-cultural studies have challenged the concept of menopause as a universal phenomenon, with wide variations in the symptom perception and reporting in women from different ethnic origins living in different countries (Avis, 2001). Cultural explanations of these differences need to include lifestyle (diet, exercise, social factors) as well as reproductive patterns, which can affect biological processes, population differences in biology as well as beliefs and attitudes to menopause and the social status of mid-aged and older women (Lock, 2005). In other words, this is a bio-psycho-socio-cultural process, which may vary within and between cultures and change over time (Richter, 1997). Unravelling these complex influences is a challenge to researchers.

The symptoms experienced by pre- and post-menopausal women are varied with little agreement about their nature, cause or severity (Mitchell *et al.*, 1996; Obermeyer, 2000). In recent years, research on menopausal symptomatology has focused on identifying symptom groupings experienced by women as they progress from pre-menopausal to post-menopausal status. However, most of these studies have been conducted among Caucasian women from western cultures. The socio-demographic characteristics, lifestyle and concomitant health problems appear to be important modifiable determinants for menopause-related symptoms. Studies (Boulet *et al.*, 1994; Dennerstein *et al.*, 1994; Guthrie *et al.*, 1994; Dennerstein, 1996; Lock, 2002) indicated that Asian women's experience of menopause is different and it affects them differently. In youth-oriented Western cultures, menopause is frequently entangled with fear of aging and loss of sexuality and even seen as a major negative life event of the same magnitude as, for example, loss of one's job or spouse (Hammond *et al.*, 1976; Bell, 1990). However, women have been reported to welcome menopause in developing countries like India, Thailand, Zimbabwe and Israel (Flint *et al.*, 1990; Sharma *et al.*, 1993; Blumberg *et al.*, 1996; MacMaster *et al.*, 1997; Punyahotra *et al.*, 1997; Singh *et al.*, 2000). Similar findings have also been reported from some developed countries (Skulton, 1970; Rosemeier, 2000). Therefore, it is difficult to extend the findings from some studies to women of other ethnic groups or cultures.

Variation in response to menopause varies within India also. A study conducted two decades ago among urban Indian women by Sharma and Saxena (1981) showed only 10% affirmative response to a direct question about their menopausal problems, with most not agreeing to have menopausal problems even if they were suffering from symptoms. Even the perception of aged Indian women immigrants did not change appreciably where they remained inside their cultural or ideological setting (George, 1988; Gupta *et al.*, 2006). They showed greater similarity with women staying in India than with women of the country of their residence.

The present paper attempted to ascertain the effect of attainment of menopause in terms of women's subjective feelings, perceived impact on femininity and availability of family support for coping with the changes.

METHOD

Menopause transcends all borders regardless of socio-economic and cultural background and, therefore, an accessible group of women from the general population was selected. The 156 participating respondents were recruited from a circle of friends, relatives and neighbours who resided in urban city of Pune, Maharashtra and occupations included that of housewife, teacher, nurse, administrative services, etc. The chief criterion for inclusion in the sample was that women be between 40 to 60 years of age and could recall what it felt like during menopause for them. A mix of primarily qualitative research with some quantitative input was carried out on a purposively selected sample of 156 women. The tools of the investigation comprised of interview guide and semi-structured schedule. Pre-tested schedule was used for the first phase of data collection. The schedule was prepared in a local language – Marathi. Words and phrases used were locally understandable and acceptable. In-depth case studies of forty-four respondents, who have had experienced the complete cessation of menses in the last five years, were taken in the second phase. It helped in gaining holistic and detailed information (with relatively less recall bias) on individual experiences, subjective description of the severity of the symptoms, felt impact on femininity, communication with and support available from spouse and other family members, coping strategies adopted, attitude towards menopause, etc.

RESULTS

Socio-economic Background

The study was conducted among women in the 40 to 60 years of age. Though the women were evenly distributed in all age groups, a large proportion (29%) of them was in the range of 50-54 years. The median age of the respondents was 50 years (SD 5.77yrs). Nearly 70% of the respondents were studied up to graduation or more. The majority (67%) of the respondents were housewives and remaining (33%) were working as professionals, service or were self-employed. One-third of the respondents were working and contributing to the family income.

Table No. 1
Characteristics of the Respondents

Age Group	*n*	%
40-44 years	36	23
45-49 years	37	24
50-54 years	46	29
55–60 years	37	24

Education		
Undergraduates	51	32.7
Graduates	70	44.8
Post-graduates	35	22.5
Marital Status		
Married	140	89.8
Not married	16	10.2
Occupation		
Housewives	104	67
Service	34	22
Professional/Self-employed	18	11
Family Type		
Nuclear	117	75
Extended	39	25
Monthly Family Income		
No response	4	2.5
Up to 10 thousand	58	37
10 to 20 thousand	53	34
> 20 thousand	41	26.5

AGE AT MENOPAUSE

Almost 60% (93) of the respondents were in post-menopausal stage and the majority (76) of them have had experienced it naturally. The mean age at menopause was 45.8 years (SD+/-4.3years). Respondents believed that the age at menopause gets influenced by family history (80), urban lifestyle (43) and genetic make up (27). Seventeen respondents reported surgical menopause. They had undergone hysterectomy or oopherectomy due to ovarian cyst, tumor, polymenorrhea, etc. Surprisingly, the majority of these respondents underwent surgery very early, between 25 to 35 years of age. A great majority (74.4%) of the respondents were aware about the symptoms and changes during menopause. Friends, colleagues and relatives were the chief sources of information. Magazines and media proved to be informative for 14% of them.

EXPERIENCE OF MENOPAUSE: SYMPTOMATOLOGY

A wide range of symptoms was attributed to menopause, as seen from Graph 1. It was difficult to distinguish between symptoms that result from loss of ovarian function, hormonal imbalance and those stemming from aging process and socio-environmental stresses of the mid-life years.

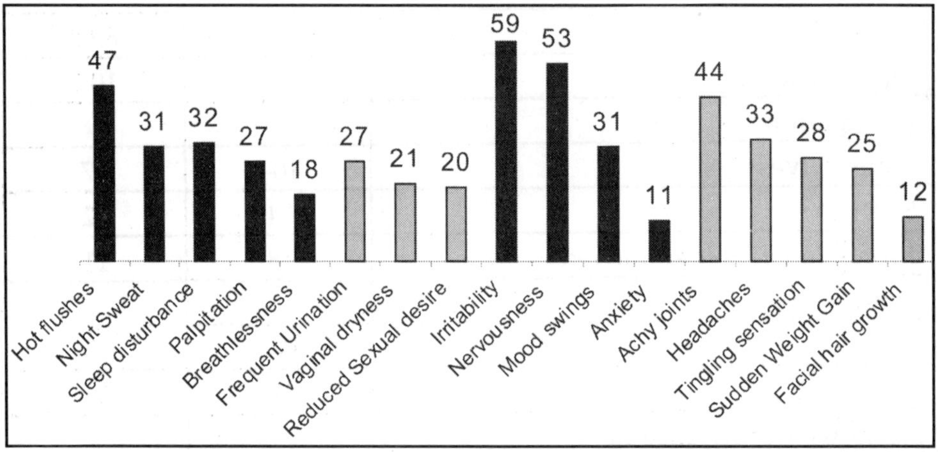

Graph 1. Frequency of Symptoms of Menopause

Of the total 156 respondents, 123 (79%) reported one or the other kind of symptoms irrespective of their menopausal state. Vasomotor symptoms like hot flushes and night sweating were accompanied also by insomnia, palpitation and breathlessness, which were reported by 49% of them. Presence of these problems was reported three years prior to the complete cessation of menses. Response to the questions on uro-genital problems was very low; only 31% answered the question. They noted frequent urination, vaginal dryness and reduced sexual desire. Irritability, nervousness and mood swings were frequently (every third woman) reported. Reporting of the psychological symptoms was more frequent than any other type of symptoms. As revealed from the data, irritability was experienced three to five years prior to menopause. Bouts of nervousness were more frequent, especially a year before cessation of menses. Achy joints, headache and tingling sensation were mentioned very frequently (47%) as compared to other symptoms like facial hair growth and sudden weight gain. Some of these problems were experienced prior to menopause. However, respondents felt that their pain and aches had aggravated because of menopause. Respondents living in nuclear families faced more psychological symptoms ($p = <0.01$) as compared to other types of symptoms. With advent of female education and changing socio-cultural milieu, women's employment in formal sector has increased but even in urban Indian setting, women have retained their traditional role of house making. Staying in nuclear family entails managing household without much help from other relatives. In case of employed respondents, stress at workplace and household management put extra burden on their emotional health. Reporting of symptoms was also influenced by income level. Women from middle and low-income groups experienced less number of symptoms as compared to high-income groups ($p = < 0.00$) when controlled for the level of awareness, education and employment status.

Distressing Symptoms

Fifty-eight respondents thought that some symptoms were distressing which troubled them and affected their routine. Irritability was reported more frequently as 'distressing' symptom because it disturbed not only mental peace but also the relationship with family members and other people around. Hot flushes and sweating during night was reported to be worrying as it created physical discomfort and resulted in sleepless nights. Disturbed sleep was troublesome when it created restlessness the following day. These symptoms are believed to have resulted in lack of confidence and lack of initiative, leading to inefficiency at workplace. Pre-menopausal women mentioned reduced sexual desire and vaginal dryness more frequently as it affected their marital relation to an extent. Respondents in pre-menopausal stage also mentioned irregular or heavy menses which acted as barriers to long-distance travelling, overnight stay outside the home and participation in religious activities. Thus, their functional area was restricted due to these symptoms.

Femininity and Menopause: Attitude

Femininity was considered to be more of a behavioral and psychological aspect, whereas loss of ability to bear child was a natural biological phenomenon. Respondents disassociated themselves with these two concepts. Most of the respondents felt that femininity was more to do with their ability to perform feminine role of caring and being responsible in family matters. To be able to carry out domestic roles along with job responsibilities was considered equally feminine. Reduced beauty due to skin changes and obesity was considered as a part of ageing process. Most of them opined that loss of femininity after menopause was a mental construct of few and definitely not a reaction of masses.

First reaction after the attainment of menopause was, 'freedom from monthly ritual'. The majority of the post-menopausal respondents (76), with the exception of sixteen, 'felt relieved' because they were not hassled by monthly cycles, associated irregularities, physical discomfort and religious restrictions. As the majority (86%) of the respondents had fruitful reproductive years and marital relations, menopause was taken in a positive spirit. They associated attainment of menopause with more 'respect' and 'promotion' to the status of senior member in the society, even though it had occurred when they were in late forties or early fifties. Menopause was a relief from the monthly menstrual tension and hence most of the respondents were looking forward to attaining menopause and entering the next phase of their life. The new phase was associated with gaining a dignified position in the society, less restricted religious life and always away from tensions of menstrual periods. In India, menstruating women face various restrictions with regard to their social interaction and religious participation. Hence, menopause was looked as a beginning of the period of 'no social and religious restrictions' and they could participate more openly and freely in religious functions and social gatherings. Though not much change in the role within the family and outside the family was noticed, they felt change in their status through their own attitude towards themselves. They expected daily life to be the same as earlier except for the monthly periods and problems arising due to the monthly cycle. Especially for the working women, menopause was a relief from the monthly trouble of periods. The psychological state of fulfilment filled their minds and collective expression of that was seen from their reactions.

At the same time, a few (16%) expressed the feeling of 'emptiness' (vacuum). They felt that 'something is missing' from life. Loss of hormonal balance was considered equivalent to loss of feminine character, especially because of changes in the reproductive physiology. Complete cessation of menstrual cycle brought a feeling of lack of something (*kamtarta aali*) or had a feeling of emptiness (*pokli nirman zali*). They believed that monthly menstrual bleeding would clean their body of 'dirty blood'. Since this process has stopped, they believed that they have grown physically weak and because of that experienced loss of interest and enthusiasm. They felt that their importance was decreased (*kamipana yeto*), (*mahatva kami hote*) and that they were of no use (*nikami janevate*). This was an indirect remark about reduced sexuality and attractiveness. Thus, they associated menopause with loss of femininity (*stritvaver parinam hoto*). This was further thought to be interfering with their confidence, self-image and self-esteem. Decreased physical capacity, loss of interest in social life, loss of sexual interest, loss of stamina, feeling of being old and feeling depressed about being of no use were associated with menopause. Surprisingly, the majority of these women were housewives and belong to the higher economic group. No definite association is found statistically though.

Spouse and Family Support

The study began with the assumption that support available to women depends on the severity of the problems faced. It was revealed that 62% of the respondents discussed their physical as well as psychological problems with their husbands. Nearly 30% mentioned that their spouse reacted positively, which means being aware of the condition, sympathetic and caring enough to offer domestic and other help. However, other family members were least considerate and did not provide any assistance to majority of them. Yet, it was also mentioned that presence of a 'grown-up' daughter provided them a supportive environment to discuss their matters. Respondents who reported more than ten symptoms relied on modern medicine. Only four respondents were on hormone replacement therapy (HRT). The rest relied on alternative therapies to modern medicines.

DISCUSSION

With the accumulation of knowledge on a woman's endocrine system, views about menopause as a natural transition began to change. According to much of current medical opinion on the subject, middle-aged women are said to suffer from an estrogen deficiency disease from menopause onwards (Utian, 1976; Wren *et al.*, 1994). In recent decades, experts from fields such as sociology, nursing and anthropology have investigated menopause and contradicted notions of menopausal women as ailing and diseased (Formanek, 1990; MacPherson, 1990; Estok *et al.*, 1991). Support is gaining for the view that a woman's menopause should not be seen as a pathologic endocrine deficiency disease because female hormones normally abate with advancing age as reproductive function comes to a halt (McKinlay *et al.*, 1992).

Similar views can be expressed through this paper as it explores more about support and perceived impact of menopause. It was seen that women were more likely to attribute certain physical symptoms to menopause, for example, aches and pains, breathlessness and weight gain, which might have been the sign of other health problems rather than menopause. The differences in attribution seen in the western and Indian groups could be due to their understanding and beliefs about menopause, but may also be due to the differences in health education between these countries.

Attitude towards menopause, views on femininity, experience of psychological changes and changes in emotional health attributed to menopause by the study population can be put into a theory suggested by Crawford. In Robert Crawford's words, the body is a cultural object. As our most immediate natural symbol, it provides us with a powerful medium through which we interpret and give expression to our individual and social experience. This expression is always mediated by culture and social roles. It defines us in relation to others in kinship, sex, age groups and larger social units such as race or caste. Bodily states are key markers in which are invested the social definitions of the self – not only regarding role, but normality and abnormality. The body also supplies a universally experienced model of a living and dynamic unit, an organic whole, a prototype from which we can draw in our attempts to explain and give meaning to larger social units and experiences. It is our richest source for metaphor (Crawford, 1984: 60-61). Similarly, the social construction of menopause in Indian set up provided a sense of relief to the respondents. As it is also considered as an entry point to old age, it was interpreted in more positive manner as one gains status and authority within the traditional age stratified Indian society.

Contradictory to this is a western thinking that menopause represents a challenging and difficult phase because, while women may feel rather young and full of energy, society tends to perceive them as becoming increasingly less attractive and less fully functioning. Especially in western societies, the time of menopause is characterised by a series of losses (e.g., loss of youth, beauty, fertility, libido, health, hormones, femininity and calcium) and there are suggestions that it possibly impacts the physical and psychological well-being in mid-life (Brown, 1976; Kaiser, 1990; Buck & Gottlieb, 1991; Bachmann, 1994; Howard & Kelly, 1994). Results showed that the crossing of this imagined portico into menopause and old age was discouraging. These women could cope with physical discomforts, even if they were severe, but encountered problems accepting aging bodies, and importantly loss of libido.

The 'loss' of fertility or decreasing interest in sexual life represented only a minor pre-occupation as most women had lived through the experience of birth, mothering and beyond and they did not wish to extend this role. Women often described feeling useless, no longer needed or wanted, not being able to sleep or rest, insecure, depressed, sexually undesirable and not appreciated and valued for their capabilities and wisdom.

With the exception of a few, respondents did not associate femininity with menopause and opined that a woman remains as feminine as before. They mentioned,

> Femininity is a mental construction while menopause is a natural process. There is an end of only reproductive capacity, which is anyways not required at this later age when the desired family size is achieved. There is a transition in role as from merely wife and mother to mother-in-law and after a few years grand mother. There is decrease in attractiveness as compared to at age of 20, but at this stage one achieves a respectful look with higher dignity as a senior member in the society.

A woman's attitude of relief from the monthly cycle was directly associated with stresses and the religious and cultural restrictions faced during menstruation. A cultural base of strong social bonding has, nonetheless, given a higher status to the elderly and thus respondents associated menopause with the grand parenting and changed (elevated) social status. The cultural context made suffering and pain of menopause easier and provided positive outlook in the minds of the respondents about mid-life changes.

Anthropological literature provides enough evidence that status and social role are important determinant of self-image and subsequent reaction, attitude and behaviour in a given social milieu. The majority were of the opinion that loss of femininity and experience of distressing symptoms was considered as a part of individuals own circumstances and no generalization can be made from a few women's experiences. Thus, the study suggested that it is necessary to look at menopause from a different perspective. It is all the more necessary to understand the role of social, cultural and behavioural factors than from mere health or clinical management perspective. Strategies used in other ethnic groups or countries may not be directly applicable to the Indian setting. This also prompts us to reconsider and decide the use of Hormone Replacement Therapy in our country.

CONCLUSION

Menopause is part of a woman's aging process. Discomfort due to physical and psychological problems are due to failure to understand the process of hormonal changes. Further menopausal processes are inextricably linked to the socio-cultural environment. While some emphasise on remaining young and beautiful, majority post-menopausal women reported the most positive attitudes toward life. Women may subscribe to bio-medical measures dealing with menopause because they have not been enabled to access different sources of information. It is exactly this situation which needs to be brought to the attention of women's health specialists – that menopause is just another passage of life women have to go through.

BIBLIOGRAPHY

Avis, N.E., R. Stellato, S. Crawford, J. Bromberger, P. Ganz, V. Cain and M. Kagawa Singer. 2001. 'Is there a menopausal syndrome? Menopausal status and symptoms across racial/ethnic groups'. *Soc Sci Med.* 52(3): 345-56.

Bachmann, G.A. 1994. 'The Changes Before "The Change": Strategies for the Transition to the Menopause'. *Postgrad Med.* 95 (4): 113-124.

Bell, S.E. 1990. 'Sociological perspectives on the medicalization of menopause'. *Ann N Y Acad Sci* 592:173-78.

Blumberg, G., B. Kaplan, D. Rabinerson, G.A. Goldman, E. Kitai and A. Neri. 1996. 'Women's attitudes towards menopause and hormone replacement therapy'. *Int J Gynecol Obstet* 54: 271-7.

Boulet, M.J., B.J. Oddens, P. Lehert, H.M. Vemer and A. Visser. 1994. 'Climacteric & Menopause in the Seven South East Asian Countries'. *Maturitas*, 19: 157-76.

Brown, M.C. 1976. 'Emotional Response to the Menopause', in S. Campbell (ed.), *The Management of the Menopause and Post-menopausal Years*. Lancaster: MTP Press, pp. 109-115.

Buck, M.M. and L.N. Gottlieb. 1991. 'The Meaning of Time: Mohawk Women at Midlife'. *Health Care Women Int.* 12(1): 41-50.

Crawford, R. 1984. 'A Cultural Account of "Health": Control, Release, and the Social Body', in J.B. McKinlay (ed.), *Issues in the political economy of health care*. New York: Tavistock, pp. 60-103.

Dennerstein, L., A.M. Smith and C. Morse. 1994. 'Psychological well-being, mid-life and the menopause'. *Maturitas* 20(1): 1-11.

Dennerstein, L. 1996. 'Well being, symptoms and the menopause transition'. *Maturitas* 23: 147-157.

Estok, P.J. and R. O'Toole. 1991. 'The Meanings of Menopause'. *Health Care Women Int.* 12(1): 27-39.

Flint, M. and R.S. Samil. 1990. 'Societal impact on menopause. Cultural and subcultural meanings of the menopause'. Part III. *Ann N Y Acad. Sci.* 592: 134-48.

Formanek, R. 1990. 'Menopause: Two Views', in M. Flint, F. Kronenberg and W. Utian (eds.), *Multidisciplinary Perspectives on Menopause*. New York: The New York Academy of Sciences, p. 418.

George, T. 1988. 'Menopause: some interpretations of the results of a study among a non-Western group'. *Maturitas* 10: 109-16.

Gupta, P., D.W. Sturdee and M.S. Hunter. 2006. 'Mid-age health in women from the Indian subcontinent (MAHWIS): General health and the experience of menopause in women'. *Climacteric* 9:13–22.

Guthrie, J.R., A.M. Smith, L. Dennerstein and C. Morse. 1994. 'Physical activity and the menopause experience: A cross-sectional study'. *Maturitas*. 20(2-3): 71-80.

Hammond, D. and A. Jablow. 1976. *Women in Cultures of the World*. London: Cummings Publishing Company.

Howard, G. and P. Kelly. 1994. 'Osteoporosis: Strategies for Prevention and Treatment'. *Mod Med Aust*. 37 (7): 18-27.

Kaiser, K. 1990. 'Cross-Cultural Perspectives on Menopause', in M. Flint, F. Kronenberg and W. Utian (eds.), *Multidisciplinary Perspectives on Menopause*. New York: The New York Academy of Sciences, pp. 430-432.

Lock, M. 2002. 'Symptom reporting at menopause: A review of cross-cultural findings'. *J Br Menopause Soc*. 8(4):132-6.

Lock, M. 2005. 'Cross cultural vasomotor symptoms reporting: conceptual and methodological issues'. *Menopause* 12: 239-41.

McMaster, J., M. Pitts and G. Poyah. 1997. 'The menopausal experiences of women in developing country: "There is a time for everything: to be a teenager, a mother and a granny"'. *Women and Health* 26: 1-13.

MacPherson, K.I. 1990. 'Nurse-Researchers Respond to the Medicalization of Menopause', in M. Flint, F. Kronenberg and W. Utian (eds.), *Multidisciplinary Perspectives on Menopause*. New York: The New York Academy of Sciences, pp. 180-184.

McKinlay, S.M., D.J. Brambilla and J.G. Posner. 1992. 'The Normal Menopause Transition'. *Maturitas* 14: 103-115.

Mitchell, E.S. and N.F. Woods. 1996. 'Symptom experiences of midlife women: Observations from the Seattle Midlife Women's Health Study'. *Maturitas* 25(1): 1-10.

Obermeyer, C.M. 2000. 'Menopause Across Cultures: A Review of the Evidence'. *Menopause* 7: 184-192.

Punyahotra, S., L. Dennerstein and P. Lehert. 1997. 'Menopausal experiences of Thai women: Symptoms and their correlates'. Part I. *Maturitas* 26: 1-7.

Richters, J.M.A. 1997. 'Menopause in different cultures'. *J Psychosom Obstet Gynaecol* 18: 73-80.

Rosemeier, H.P. 2000. 'Turning points in a woman's life – a psychosocial perspective from the menarche to the menopause in European women'. *Adv Obstet Gynecol* 52: 217-25.

Sharma, V.K. and M. Saxena. 1993. 'Menopausal symptomatology: its influence on the attitude and subjective adaptation level of the middle aged climacteric (Indian) women', in C.M. Agarawal (ed.), *Dimensions of Indian Womanhood*. Vol. 1, First edn., Almora Book Depot, pp. 317-36.

Sharma, V.K. and M.S. Saxena. 1981. 'Climacteric symptoms: a study in the Indian context'. *Maturitas* 3: 11-20.

Sengupta, A. 2003. 'The emergence of the menopause in India'. *Climacteric* 6: 92-95.

Singh, A. and A.K. Arora. 2000. 'Menopausal women's profile in rural North India – an integrated qualitative and qualitative study'. *Adv Obstet Gynecol* 52: 309-13.

Skultans, W.E. 1970. 'The symbolic significance of menstruation and menopause'. *Man* 5: 639-51.

Utian, W.H. 1976. 'Scientific Basis for Post-Menopausal Estrogen Therapy: The Management of Specific Symptoms and Rationale for Long-Term Replacement', in R.J. Beard (ed.), *The Menopause: A Guide to Current Research*. Lancaster: MTP Press, pp. 175-201.

Wren, B.G. and J.A. Eden. 1994. 'Hormone Replacement Therapy: A Review - Part I', in *The Female Patient*. Hong Kong: Excerpta Medica, pp. 5-18.

16

FINGER DERMATOGLYPHIC VARIATION AMONG MENTALLY CHALLENGED CHILDREN OF VISAKHAPATNAM

Shibani Mukhopadhyay and S. Narahari

INTRODUCTION

Dermatoglyphic data, besides population variation (Mukherjee, 1976; Malhotra, 1978), have been demonstrated as a diagnostic tool in clinical genetics particularly in chromosomal aberrations (Uchida, 1967). Turpin and Lejeune (1953), Penrose (1954, 1963, 1968), Loesch (1983, 1986), Banerjee *et al.* (1991) and Shanti Devi (1992) reported the significance of dermatoglyphics in chromosomal aberrations, i.e., mentally disabled/ Down's Syndrome. Cummins and Midlo (1961) listed some of the references on dermatoglyphics related to clinical and medical genetics. Recently, Mukherjee (2006) focused the clinical importance of td ridge count on human palms. The present paper deals with Finger Dermatoglyphic Variation among the mentally disabled children of Visakhapatnam, Andhra Pradesh, hitherto unreported.

To define mental disability or mental challenge aptly is an intricate process because it has legal status and deserves a clear and reliable definition (Ingalis, 1978). It is defined variedly on the basis of intelligence test score, failure in social adjustment and essential causes such as brain damage. On the basis of intelligence test score, it is defined as any person with an IQ below a certain level, usually set at 70. On the basis of social adjustment, "A mentally defective person is a person who is incapable of managing himself and his affairs, or being taught to do so and who requires supervision, control and care for his own welfare and welfare of the community" (Benda, 1954). On the basis of causes, it is defined as the children who suffered from a severe brain disease either in uterus or in early childhood that disturbed the normal development of the brain and produced serious anomalies in mental development. According to the American Association of Mental Deficiency (AAMD), "Mental retardation refers to significantly sub-average general intellectual functioning existing concurrently with deficits in adaptive behaviour and manifested during the development period"(Grossman, 1973). In other words, a person who suffers from an impairment in his ability to think, learn and reason is called mentally disabled. On the basis of IQ, there are 4 levels of mental disability, viz. mild, Moderate, Severe and Profound. Regarding the prevalence of mental disability in human populations, there is no uniformity of its occurrence in different segments and sex. About 3% of world population shows mental disability (Bartam, 1975). In India, an estimate of 20 million persons of mild disability and about 4 million with moderate and severe level are evinced. The present paper, due to inadequate sample size for each level of MD, confines to the single category, i.e., mentally disabled or mentally challenged.

MATERIAL AND METHODS

Bilateral inked finger ball prints were collected from a total of 315 mentally challenged (MC) from 7 special schools, viz., BHPV Special School, Gajuwaka, Bethany SP School, Siripuram, Pimhans SP School, Akkayyapalem, Sankalp, Malkapuram, SR Puram SP School, Srirampuam, Priyadarhini Integrated School, Isukathota, and Lebenshilpe, MVP, Colony of Visakhapatnam district. 315 normal/control children were selected keeping in view age, sex and social group aspects, from the general population of Visakhapatnam. The age of the subjects ranges from 6+ to 16+ years. The prints were analysed for both qualitative and quantitative characters. The qualitative characters include Finger Pattern Types, while that of quantitative composed of Finger Pattern Intensity Index (FPII) and Total and Absolute Ridge Counts (TFRC and AFRC). The suggestions of Cummins and Midlo (1961), Mukherjee (1967, 1974) and Holt (1968) were followed in the collection and analyses of the data. The tests of significance, i.e., Chi-Square and t-test were computed wherever necessary, to evaluate intra- and inter-group differences.

RESULTS AND DISCUSSION

The frequency distribution of Finger Pattern Types and mean values of Finger Pattern Intensity Index (FPII), Total and Absolute Ridge Counts (TFRC & AFRC) among the mentally challenged (MC) and normal children

(Table 1) depicts a preponderance of whorls (38.63%) and arches (2.32%) in MC than normal children who show a high frequency of loops (72.67%). Females possess an excess value for whorls in MC, whereas it is observed in the normal for loops and whorls compared to their counterparts. The sex difference is statistically significant only for the normal/control group.

Table No. 1

Finger Dermatoglyphic Characters among the Mentally Challenged and Normal Children

	Mentally Challenged			Normal		
Character	**Male**	**Female**	**Total**	**Male**	**Female**	**Total**
I. Qualitative:						
a. Finger Pattern Types(%):						
i. Arch	***02.45***	***02.09***	***02.32***	***01.61***	***00.64***	***01.27***
Plain	02.24	01.09	02.13	01.36	01.76	01.56
Tented	00.19	01.82	00.09	00.05	00.00	00.03
ii. Loop	***59.17***	***58.82***	***59.05***	***72.54***	***72.91***	***72.67***
Radial	02.78	03.48	03.02	02.73	01.82	02.41
Ulnar	56.39	55.36	56.03	69.80	71.09	70.25
iii. Whorl	***38.39***	***39.09***	***38.63***	***25.85***	***26.45***	***26.06***
True	32.15	33.91	32.76	23.32	23.54	23.40
Composites						
TL	02.24	01.45	01.97	00.78	02.27	01.30
LPL	01.23	00.82	01.08	01.07	00.36	00.82
CPL	02.78	02.91	02.82	00.68	00.27	00.54
Accidental	00.00	00.00	00.00	00.00	00.00	00.00
Chi-Square value for Sex Dimorphism: 15.68* (Significant at 5% probability)						
II. Quantitative (Mean + S.E.)						
FPII	13.57+0.25	13.71+0.31		12.44+0.20	12.57+0.25	
TFRC	112.27+2.70	112.67+3.53		86.73+1.57	87.86+2.08	
AFRC	154.97+5.18	155.13+6.75		108.20+2.87	111.58+3.96	

In Henry's Classification, the MC children show an elevated frequency of Plain Arch (2.13%) and Central Pocket Loop (2.82%) followed by Twin Loops and Lateral Pocket Loops. Radial loops. In contrast, the normal children show an excess of Plain Arches (1.56%), Twin Loops followed by Lateral Pocket Loops (LPL) and Central Pocket Loops (CPL). The sex-wise distribution indicates greater variation (about 6%) in whorls for Henry's Classification between male and female MC children. The sex difference is statistically significant for both MC and normal children, which corroborates with Mavalwala (1978) and Bhanu's (1991) contention that micro-level details of basic pattern types show greater variation.

In Quantitative Characters, the MC children show higher mean values of the variables than normal children. The sex-wise trend indicates that females depict relatively higher values than males for all the variables in both MC and normal children. However, none of the variables reveal statistically significant sex dimorphism in either group under study.

Thus, the finger dermatoglyphic data in general projects higher values for both qualitative and quantitative characters in MC children than normal. The significant sex dimorphism noted for the qualitative than quantitative characters signifies patterns in terms of shape rather than size depict greater variation in the MC compared to normal children. The occurrence of higher frequency of finger dermatoglyphic characters and non-significant sex difference for quantitative characters among the MC children is in congruence with the findings of Shanti Devi (1992) but deviates from Rajangam *et al.* (1991) and Banerjee *et al.* (1991), who reported lower values in Down's Syndrome patients than normal ones.

ACKNOWLEDGEMENTS

The authors express their deep sense of gratitude to all the faculty members of the special schools who extended their cooperation and help during data collection on mentally disabled or challenged children.

BIBLIOGRAPHY

Banerjee, A.R., S. Banerjee, A.R. Bandopadhyay, S.C. Paul and S.P. Banerjee. 1991. 'A note on the dermatoglyphic features of Down Patients', in B. Mohan Reddy, S.B. Roy and B.N. Sarkar (eds.), *Dermatoglyphics Today*. Calcutta: IBRAD, pp. 206-209.

Bartam. 1975. Cited from Shanti Devi, 1992.

Benda. 1954. Cited from Shanti Devi, 1992.

Bhanu, B.V. 1991. 'Digital Patterns: A New Classification System', in B. Mohan Reddy, S.B. Roy and B.N. Sarkar (eds.), *Dermatoglyphics Today*. Kolkata: IBRAD, pp. 1-16.

Cummins, H. and C. Midlo. 1961. *Finger Prints, Palms and Sole*. An Introduction to Dermatoglyphics. New York: Dover.

Grossman, H.J. (ed.). 1973. *Manual on terminology and classification in mental retardation*. Washington DC: American Association of Mental Deficiency.

Holt, Sarah B. 1968. *The Genetics of Dermal Ridges*. Springfield: CC Thomas.

Ingalis. 1978. *Mental Retardation: The Changing outlook*. New York: Dept. of Mental Hygiene.

Loesch, D.Z. 1983. *Quantitative Dermatoglyphics: Classification, Genetics and Pathology*. Oxford: Oxford University Press.

——. 1986. 'Dermatoglyphic findings in Fragile X syndrome: A causal hypothesis points to X-Y interchange'. *Ann. Hum. Genet*. 50: 385-398.

Malhotra, K.C. 1978. 'Status of Dermatoglyphics Studies in India: A Comment', *SAP* 1: 1-8.

Mavalwala, J. 1978. 'A method for dermatoglyphics: fingers and palms', in J. Mavalwala (ed.), *Dermatoglyphics: An International Perspective*. The Hague: Mouton, pp. 19-54.

Mukherjee, D.P. 1967. 'Quantitative Genetics of Dermal Ridges with special reference to an Indian Population'. Ph.D. Thesis, London University.

——. 1974. 'New Approaches to Dermatoglyphics Methodology', in L.D. Sanghvi, V. Balakrishnan, H.M. Bhatia, P.K. Sukumaran and Undevia (eds.), *Human Population Genetics in India*. New Delhi: Orient Longman Ltd.

——. 1976. 'Trends in Dermatoglyphic Research', in H.K. Rakshit (ed.), *Anthropology in India. 2. Physical Anthropology*. Calcutta: Anthropological Survey of India.

——. 2006. 'Anthropology and Clinical Importance of td Ridge Count on Human Palms'. *The Anthropologist* 8(4): 219-222.

Penrose, L.S. 1954. 'The distal triradius on the hands of parents and sibs of Mongol imbeciles'. *Ann. Hum. Genet.* 19: 10-38.

——. 1963. 'Finger Prints, Palms and Chromosomes'. *Nature* 197: 709-712.

——. 1968. 'Memorandum on dermatoglyphic nomenclature'. *Birth Defects. Original Article Series*, 4: 1-12.

Rajangam, S.S. Janakiram and J.M. Thomas. 1991. 'Dermatoglyphics in Indian Downs and their parents', in B. Mohan Reddy, S.B. Roy and B.N. Sarkar (eds.), *Dermatoglyphics Today*. Calcutta: IBRAD.

Shanti Devi. 1992. 'Genetic Studies in Mentally Retarded Children'. Ph.D. Thesis, Vishakhapatnam: Andhra University.

Turpin and Lejeune. 1953. Cited from Penrose, 1963.

Uchida, T.A. 1967. 'Chromosome aberrations induced in vitro by low doses of radiation; non-disjunction in lymphocytes of young adults'. *Am. J. Hum. Genet.* 27: 419-429.

17

IMPACT OF SMOKING, ALCOHOLISM AND CONSANGUINITY ON REPRODUCTIVE MORTALITY: THE GADABA CASE

S. Narahari

INTRODUCTION

Fertility and mortality are the two ends of the scale of demographic status of a population. They are influenced by several factors including socio-cultural and biological, viz., habits like smoking, tobacco chewing and alcoholism, and mating structure, i.e., consanguinity and blood group incompatibility. The effect of consanguinity on fertility and mortality was well explored in different populations which states that high fertility and mortality are evident in consanguineous than non-consanguineous unions (See Narahari and Aparna, 2005). Sengupta and Purnungle (2004) showed that early menarche and marriage age in consanguineous unions leads to elevate reproductive mortality, i.e., still-births and abortions. Blanco Villega and Fuster (2006) observed in population of Spain that consanguinity prolong reproductive period paving for greater number of offsprings and lower infertility and possible compensate effect for infant mortality. The impact of smoking was reported causing delay in conception, ectopic pregnancy, improper nourishment of fetus through placenta leading to reproductive mortality, congenital malformations and advancement of menopause by two years (Midgette and Baron, 1990; Poswillo and Alberman, 1992; Narahari and Rani, 2002; Polanska and Hanka, 2004; Nakamura *et al.*, 2004, Venners *et al.*, 2004; Khoury *et al.*, 2004). Maternal alcoholism is also known to have adverse effect on reproductive performance and fetal development (Larroque, 1992) though the effect of moderate consumption remain controversial. Kesmodel *et al.* (2002) have reported an increased risk involvement of spontaneous abortions in mothers of moderate consumption. Thus, it is clear that there exists hardly any report on the combined effect of habits and consanguinity on reproductive performance excepting the study of Aparna and Narahari (2007) that focused the impact of both smoking and consanguinity together on reproductive performance in a fishing community, viz., Vadabalija of Vizianagaram district of Andhra Pradesh. Therefore, the present study has been undertaken to investigate the combined influence of habits and consanguinity on fertility (pregnancies/ live births) and reproductive mortality or reproductive wastage or pregnancy wastage (still-births and abortions) in a Primitive Tribal Population of Andhra Pradesh, viz., the Gadaba.

The state of Andhra Pradesh in South India is well known and stands unique for high frequency of consanguinity and inbreeding levels (Sanghvi, 1966; Basu,1971; Reid, 1971 & 1973; Roy Choudhury, 1976) among different tribal and non-tribal populations who differ widely in ethnicity, habitat, socio-cultural practices, economy, language and so on. The estimated tribal population of the state is 4.2 millions. There are eight Primitive Tribal Groups (PTGs) in the state (Dash Sharma, 2006), of which four are inhabiting the North-Coastal Districts (Srikakulam, Vizianagaram and Vishakhapatnam), viz., the Savara, the Gadaba (present study), the Porja and the Khond.

The Gadabas are mainly distributed in the fringe districts of Andhra Pradesh and Orissa. According to Census Report 1971, they number 25,108 of which 12,408 were men and 12,700 women while Census 1981 shows an increase of 2,624 in population size. The people speak their own dialect which belongs to Mundari group (Majumdar, 1941; Aiyappan, 1948; and Somasundaram, 1948). They are cultivators (settled and *podu*/shifting) and also engage themselves in gathering minor forest produce. Some of them have small land-holdings. The social structure reveals that the Gadaba is a large endogamous group which consists of several sub-divisions. Aiyappan (1948) mentioned six sub-divisions while Thurston (1909) observed five sections. In Vizianagaram district during the field-work, the people say that there are four major social divisions, viz., Kathri Gadaba, Thudam Gadaba, Parangi or Goli Gadaba and Olari Gadaba on the basis of social customs and religious practices. There is Kancham Sambandam or inter-dining relationship among these sub-divisions (Ravi, 2003). The people prefer and practise consanguineous marriages. The habit of smoking, particularly reverse smoking or *Adda Poga,* is in practice among a great majority of men and women. However, a few of them also consume alcohol. Hence, the present study has been undertaken on this population.

MATERIAL AND METHODS

Relevant data on demographic events such as pregnancies, live births, surviving offspring, still-births and abortions, post-natal mortality and habits like smoking and alcoholism were collected using Schedule-cum-Interview method from a total of 388 ever married women, who have experienced at least one pregnancy in their lifetime. The sample hails from 17 villages belonging to 2 Mandals, viz., Pachipenta and Salur of Vizianagaram district of Andhra Pradesh. Women who smoke at least 2 cigars per day were considered as smokers and who consume beverage (indigenous/commercial) at least 3 times per week as alcoholic. At the same time, those women who are totally devoid of the practice of smoking and alcohol consumption were considered as controls for comparison. Detailed pedigrees were constructed to evolve the pattern and prevalence of consanguinity and in-breeding level. The frequency distribution of women in accordance with smoking, alcoholism and consanguinity at Mandal level was given in Table No. 1. It is evident that about 76% of smokers, 5% of alcoholic and 52% of consanguinity are found in Gadaba women.

Table No. 1

Distribution of Gadaba Women by Consanguinity, Smoking and Alcoholism in Two Mandals of Vizianagaram District

S.No.	Category	Pachipenta Mandal (9 Villages)	Salur Mandal (8 Villages)	Total
1. a	Consanguineous	119	84	203
b	Non-consanguineous	94	91	185
	Total	213	175	388
2. a	Smoking	154	139	293
b	Non-smoking	59	36	95
	Total	213	175	388
3. a	Alcoholic	05	16	21
b	Non-alcoholic	208	159	367
	Total	213	175	388

RESULTS AND DISCUSSION

The mean values of fertility and reproductive mortality (still-births and abortions) among the Gadaba women with habits and consanguinity (Table 2) relatively show higher values for pregnancy and reproductive mortality among the women with smoking and consanguineous compared

Table No. 2
Habits, Consanguinity and Reproduction among the Gadabas

S. No.	Category	Pregnancies			Still-births			Abortions			Total RM		
		No	Mean	t-Value	No	Mean	t-Value	No	Mean	t-Value	No	Mean	t-Value
1.	Smoking	1041	3.64	**3.77****	79	0.28	0.44	65	0.23	1.37	144	0.50	1.08
2.	Non-smoking	275	2.93		23	0.24		15	0.16		38	0.40	
3.	Alcoholic	66	3.14	0.93	02	0.10	1.37	04	0.19	0.23	06	0.29	1.18
4.	Non-alcoholic	1250	3.48		100	0.28		76	0.21		176	0.49	
5.	Consang-uineous	731	3.66	**2.45***	51	0.61	0.46	43	0.22	0.22	94	0.47	0.24
6.	Non-consang-uineous	585	3.25		51	0.28		37	0.21		88	0.49	

to the women with non-smoking and non-consanguineous, but significant group difference is evident only for pregnancies. On the other hand, women with alcoholism show slightly decreased mean values for both pregnancy and reproductive mortality compared to women without alcoholism. Yet, the difference is statistically non-significant.

The combinational impact of smoking, alcoholism and consanguinity on reproductive performance (Table No. 3) reveals that there is an elevated trend in fertility (pregnancies/live births) from consanguineous to non-consanguineous and smoking to non-smoking categories while a decreased trend of it for alcoholism to non-alcoholism category.

Table No. 3

Smoking, Alcoholism and Consanguinity in Relation to Fertility and Reproductive Mortality among the Gadabas

Group	Pregnancies N Mean	Live Births N Mean	Still Births N Mean	Abortions N Mean	Repro-ductive Mortality N Mean
Smoking consanguineous	587 3.86	514 3.38	38 0.25	35 0.23	73 0.48
Non-smoking consanguineous	144 3.00	123 2.56	13 0.27	08 0.17	21 0.44
Smoking non-consanguineous	454 3.39	383 2.86	41 0.31	30 0.22	71 0.53
Non-smoking non-consanguineous	131 2.85	114 2.48	10 0.22	07 0.15	17 0.37
Alcohol consanguineous	36 3.54	42 3.23	00 0.00	04 0.31	04 0.31
Non-alcohol consanguineous	685 3.66	595 3.18	51 0.27	39 0.21	90 0.48
Alcohol non-consanguineous	20 2.50	18 2.25	02 0.25	00 0.00	02 0.25
Non-alcohol non-consanguineous	565 3.28	479 2.78	49 0.28	37 0.22	86 0.50

The proportion of increment in fertility between the said categories is 3% and 13%, respectively with a depletion of 19% for the alcoholic. When habit and consanguinity are considered together, there is almost double the proportional increment of fertility between smoking-consanguineous and non-smoking-non-consanguineous (26%) compared to alcoholic-consanguineous and non-alcoholic-non-consanguineous (14%) groups.

Thus, the findings corroborate with those of earlier studies (See Narahari & Aparna, 2005) where consanguinity elevates fertility. In addition, the combined effect of both consanguineous and smoking attitude triggers enormously the increment. However, in case of alcoholics, it tends to decrease fertility showing a negative impact compared to their counterparts.

Reproductive mortality in relation to consanguineous and non-consanguineous, smokers and non-smokers, alcoholic and non-alcoholic women (Table No. 2) display a proportional increment of 16% between consanguineous and non-consanguineous and 30% between smokers and non-smokers but a decrease of 50% was noted between alcoholic and non-alcoholic, thereby revealing a reverse trend for alcoholism and reproductive mortality.

For the components of reproductive mortality, considered separately, the values display a great influence of consanguinity on still-births whereas it is that for smoking on abortions. Both the components express about 3% of variation of the impact of factors under study. At the same time, a negative impact is observed for alcoholic on still-births but unclear in case of abortions.

When habits and consanguinity are considered together, there appears positive affect between smoking and consanguineous and their counterparts (23%), and a negative effect between alcoholic and consanguineous and their counterparts (38%). The components of reproductive mortality further reveal that there is greater variation (about 35% and 29%), respectively, of positive impact for the abortions between smoking and consanguineous and for alcoholic and consanguineous compared to their counterparts. However, in case of abortions only positive effect is noted for smoking and consanguineous (12%), while it is exactly a reverse trend for the combination of alcoholic and consanguineous compared to their respective counterparts.

Thus, the present study clearly projects the impact of habit and consanguinity where greater variability (impact) is evident for the combination of smoking and consanguinity as well as alcoholism and consanguinity. It also corroborates with earlier studies that consanguinity elevates both fertility and mortality. In addition, it also projects that smoking increases reproductive mortality more than non-smoking. However, the effect of alcoholism seems to be in reverse direction wherein it decreases fertility and mortality in alcoholic than non-alcoholic women. Therefore, the present study suggests to conduct more studies on this aspect on different populations of India besides counseling the Gadaba tribal population to avoid smoking, alcoholism and practice consanguineous marriages.

ACKNOWLEDGEMENTS

The author expresses his gratitude to the University Grants Commission, New Delhi for financial support of Major Research Project, "Habits and Reproduction: A Bio-Anthropological Study among the Primitive Tribal Groups of North-Coastal Andhra Pradesh, India" [Letter No. F.5-s174/ 2004(HRP) Dated 21-10-2004]. Thanks are due to ITDA, Parvathipuram and also the Gadaba people who extended their cooperation and help during data collection.

BIBLIOGRAPHY

Aiyappan, A. 1948. *Report on the Socio-economic Conditions of the Aboriginal Tribes of the Province of Madras*. Madras: Government Press, pp. 67-70.

Aparna, A. and S. Narahari. 2007. 'The impact of smoking and consanguinity on Reproductive Success among the Vadabalijas: A fishing community of Vizianagaram District, Andhra Pradesh (India)', in A.N. Sharma (ed.), *Health Studies in Anthropology*. New Delhi: Sarup and Sons, 142-148.

Blanco Villega, M.J. and V. Fuster. 2006. 'Reproductive Pattern in Consanguineous and Non-Consanguineous marriages in La Cabera, Spain'. *Annals of Human Biology* 33(3): 330-341.

Census of India. 1971. *Social and Cultural Tables and Special Tables on Scheduled Castes and Scheduled Tribes. Series 2. Andhra Pradesh, Part I and Part V (A)*. Delhi: Controller of Publications, 1978.

Census of India. 1981. *Special Tables for Scheduled Tribes. Series 2, Andhra Pradesh. Part IX(iii)*. Delhi: Controller of Publications, 1988.

Kesmodel, U., K. Wisberg., S.F. Oslen., T.B. Henriksen and N. Secher. 2002. 'Moderate alcohol intake in pregnancy and the risk of spontaneous abortion'. *Alcohol Alcohol* 37 (1): 87-92.

Khoury, J.C., M. Miodovaik, C.R. Buncher, H. Kaikwarf, Mc Elvys, P.R. Khoury and B. Sibai. 2004. 'Consequences of smoking and caffeine consumption during pregnancy in women with type-1 diabetes'. *J. Matern Fetal Neonatal Med.*, Jan., 15(1): 44-50.

Larroque, B. 1992. 'Alcohol and the Fetus'. *International Journal of Epidemiology* 21 Suppl.: S8-16.

Majumdar, D.N. 1941. 'Racial affiliation of the Gond of Central Provinces'. *Journal of the Royal Society of Bengal, Science* VII: 35-36.

Midgette, A. and J.A. Baron. 1990. 'Cigarette smoking and the risk of natural menopause'. *Epidemiology* 1: 474-480.

Nakamura, M.U., S.M. Alexandre, J.F. Kuhn Dos Santos, E. De Souzae, M. Sass, A.P. Auritscher Beck, E. Trayna, C.M. Andrade, T. Barroso and L. Kulay Junior. 2004. 'Obstetric and prenatal effects of active and/or passive smoking during pregnancy'. *Sao Paulo Med. J.* May 6: 122(3): 94-98.

Narahari, S. and A. Aparna. 2005. 'Consanguinity, Reproduction and Selection Potential: A Study in a Fishing Community of Vizianagaram District, Andhra Pradesh'. *The Oriental Anthropologist* 5(2): 216-220.

Narahari, S. and D.S. Rani. 2002. 'Habits and Reproductive Performance in a Food Gathering Tribe of Andhra Pradesh, India'. *Journal of Human Ecology* 13(3): 255-256.

Polanska, K. and W. Hanka. 2004. 'Influence of smoking during pregnancy on maternal condition and birth outcome: Overview of epidemiological studies'. *Przegl. Epidemiology* 58: 683-691.

Poswillo, D. and E. Alberman. 1992. 'Effects of smoking on the fetus, neonate and child'. *OUP.*

Ravi, K. 2003. 'Gadaba', in *People of India, Andhra Pradesh.* Vol. XIII, Part I. New Delhi: Anthropological Survey of India/Affliated East-West Press, pp. 588-594.

Reid, R.M. 1971. 'Marriage Patterns, Demography and Population Genetics in a South Indian Caste: A Study of Inbreeding in Human Population'. Ph.D. Thesis, Urbana, USA: University of Illionis.

——. 1973. 'Social Structure and Inbreeding in a South Indian Caste', in N.E. Morton (ed.), *Genetic Structure of Population.* University of Hawaii, p. 92.

Roy Choudhury, K. 1976. 'Inbreeding in Indian Populations'. *Trans. Bose Res.* 39: 65-76.

Sanghvi, L.D. 1966. 'Inbreeding in India'. *Eugenics Quarterly* 13: 291-301.

Sengupta, S. and Purnungle. 2004. 'Bio-Social determinants of Fertility: A case study among the Ao Nagas of Mokochung district, Nagaland'. *Vanyajati,* April Vol., 16-20.

Somasundaram, A.M. 1949. 'A Note on the Gadabas of Koraput District'. *Man in India* 29: 36-45.

Thurston, E. 1909. *Castes and Tribes of Southern India.* Part II, Madras: Government Press, pp. 242-252.

Venners, S.A., X. Wang, C. Chen, I. Wang, D. Chen, W. Guang, A. Huang, L. Ryan, J. O' Connor, B. Lasley, J. Overstreet and Xu X Wilcox A. 2004. 'Paternal smoking and pregnancy loss: a prospective study using a biomarker of pregnancy'. *American Journal of Epidemiology* May 15: 159(10): 993-1001.

18

MATERNAL HEALTH CARE UTILIZATION IN SIKKIM

Shailendra K. Mishra, Sujata Kar and Susmita Mukhopadhyay

INTRODUCTION

Throughout the world, birth of a new baby is celebrated. Societies expect women to bear children and honour them for their role as mother. Every year, more than 200 million women become pregnant and at least 15% are likely to develop complications that may require skilled obstetric care to prevent death and serious ill health (WHO, 1994). It is distressing that every year, more than half a million women die during pregnancy or childbirth (WHO, 2004). The major clinical causes of maternal deaths include postpartum haemorrhage, obstructed labour, infection, eclampsia and septic abortion. The surviving mothers might suffer from perineal tears, genital infections, uterogenital prolapse and vesico-veginal fistula (Coyaji, 1991). The immediate causes of maternal deaths mask the underlying mechanism, which can explain the possible reasons of risks. Utilization of antenatal care (ANC) and delivery care facilities can largely help to mitigate the risks of maternal mortality. Studies reveal that most of the maternal deaths could be prevented if women had received adequate and timely care services; proper prenatal care and referral to appropriate health care facilities (Jejeebhoy and Rao, 1995).

In India, maternal health has received top priority among the objectives of Family Welfare Programme. Time to time programmes have been reviewed and modified to ensure maximum effectiveness (MHFW, 1997, 1998). In spite of consistent efforts, India has a high maternal mortality ratio of approximately 450 deaths per 100,000 births (WHO, 1996). Differential maternal mortality ratio was reported for different states, which can be attributed to provision of resource settings and socio-demographic profile of the respective population. Access to health care facilities and level of health care utilization is largely related to socio-demographic factors. Moreover, making motherhood safer requires commitment from all the facets, which include home, the community and clinical facilities, etc. (Matthews *et al.*, 2001). It also involves how women themselves perceive the complications of pregnancy and how they react to those complications.

Though there is no consensus among the scholars about the method of assessment of socio-economic status (Lebiratos *et al.*, 1988), there have been several studies considering different socio-demographic parameters and their relationship with the utilization of antenatal care and safe delivery care. Most of these studies are based on urban and rural habitats. Rural residence considered as a proxy of the accessibility of professional care or distance, which showed a very strong negative effect on utilization of facilities (Pebley, 1996, Bhatia and Cleland, 1997). A study in Peru signifies the effects of maternal schooling on the use of prenatal care and delivery assistance (Elo, 1992). Urban residence, high standard of living and high maternal education influence the utilization of antenatal care (Obermeyer, 1991). Carlough (1997) found that costs, transportation, family beliefs and decision making potentially determine the timely utilization of maternal care facilities. A British study based on composite measures of clinical, provider and socio-demographic factors states social class measures as significant predictor of maternal care utilization (Petrou *et al.*, 2001). Wagle *et al.* (2004) reported accessibility to health facilities, in terms of distance, as major determinant of the place of delivery whereas another study reports discrepancies in utilization of facilities with the place of residence (Elo, 1992). A study conducted in Karnataka state of India illustrates the importance of parity and socio-economic status. The same study also recognizes antenatal care as preventive measure of infant mortality (Chandrashekar *et al.*, 1997).

In general, a number of studies demonstrate that use of ANC and skilled attendance at delivery is positively related to the economic condition, employment and educational status of mothers along with problems experienced during previous delivery, perception of risk associated to the birth and media exposure (Obermeyer and Potter, 1991; Dujardin *et al.*, 1995; Obermeyer, 1993; Robley *et al.*, 1996; Bhatia and Cleyland, 1995; Chandihok. 2006). However, negative relations were encountered with parity and distance from health facilities (Bolam *et al.*, 1997).

In developing countries, antenatal care and delivery care practices can be conceived as major determinants of maternal morbidity and mortality. WHO infers antenatal care and safe delivery care as two important pillars of safe motherhood initiative (WHO, 1994; Palaniappan, 1995). It is important to identify the risk factors, which lead to acceptance and unacceptance of professional care facilities during pregnancy and at birth.

The present study investigates the association of socio-demographic characteristics with the acceptance of antenatal care during pregnancy and use of skilled attendance at delivery among women of two groups residing in Sikkim.

MATERIALS AND METHODS

Data were collected during Dec. 2004 to Feb. 2005 as part of a large ongoing project among two ethnic groups – Rai and Bhutia – inhabiting in the same area at Ranka, about 13 km away from Gangtok, the capital town of Sikkim.

In the study area, households are sparsely distributed on small hillocks with a concentration near roads. Between two populations, numerical strength of the Rai is higher than the Bhutia. Both communities are migrants to that area. The Bhutia probably migrated to Sikkim in sixteenth century from Tibet, whereas the Rais have reported to be settled in Sikkim from seventeenth century, in the wake of the invasions of the Gorkha of Nepal. Both of the communities are agriculturists, practising settled terrace cultivation. Women play a major role in agricultural activities apart from their normal household chores.

Almost all the Rai and Bhutia households of the area were included for the study. Informed consent was sought for the participation in the study through a consent form filled by participants, prior to the study.

Data were collected through direct interviews of women who had given birth to at least one child and are still in their active reproductive age. The study includes cases of 160 only, mean age of whom is 30.38 (±7.40) years. Female interviewers, who had prior experience of conducting surveys and were familiar with the local culture, conducted interviews. Before actual operation of the survey, these interviewers underwent extensive training to ensure that they understood all the questions thoroughly and were able to administer those uniformly without hesitation. Questions were prepared and pre-tested before study to ensure easy description of terminologies that the women were able to understand.

Data were collected about socio-demographic aspects of the family and different aspects of reproductive health, which include knowledge and practices related to menstruation, pregnancy, childbirth, antenatal care, delivery care, etc.

Pearson's Chi-square test was used to assess the association of the acceptance of antenatal care and utilization of skilled attendance at delivery with the socio-demographic characteristics. Logistic regression (Agresti, 1990) was done to assess possible predictors of acceptance of antenatal care and likelihood of attended delivery. Independent variables used for logistic regression include the level of mother's education, ethnicity, occupation, parity and monthly expenditure of the family. In order to avoid the effect of confounding, only level of education has been considered as indicator of literacy of mothers. Similarly, monthly expenditure of family was included as indicator of economic status to avoid confounding with the other indicator, like monthly income. For Chi-square and logistic regression, data were pooled and ethnicity was included as an explanatory variable. Before logistic regression, Chi-square test of trend was utilized to shows the trend of a particular trait in the analysis.

Mothers have been divided into four categories in the increasing order of education, i.e., illiterate, primary, secondary and higher education level. Similarly, three groups are formed for occupation, i.e., agriculture or other labourers, housewife and mothers employed in service sector. For ethnicity, two groups, viz., Rai and Bhutia are considered. For parity mothers are divided into three groups, i.e., primipara, multipara and grand multipara with increasing order of parity one, two to four and more than four respectively.

Binomial logistic regression was fitted for antenatal care taken during pregnancy as well as for utilization of skilled attendance at delivery to estimate the effects of explanatory variables. Baseline categories for regression are antenatal care not taken during pregnancy and no utilization of skilled attendance at delivery. Reference category for level of education is illiterate, for ethnicity Rai and for mother's occupation agricultural/ other labourers. Similarly, for monthly expenditure of family, group-I ($\geq$ 2000 Rs.) and for parity, primipara have been considered as reference category. Model-I includes level of education as the only explanatory variable. Further, the data were controlled for ethnicity (Model-II), ethnicity and occupation (Model-III), ethnicity, mother's occupation and monthly expenditure of the family (Model-IV) and finally for parity along with all other variables of Model-IV (Model-V). It is done in order to observe the changes in the odds ratios (OR) for the level of education.

All statistical analysis was done with the help of SPSS (7.5) and Medcalc statistical package (version 7.2.0.2, Frank Schoonjans, Marikerke, Belgium).

RESULTS

Table No. 1 shows the socio-demographic characteristics of both communities. It shows that about 67% Rai and 56% Bhutia women have attained at least primary level of education. The major proportion of women (60.2% of Rai and 66.7% of Bhutia) is engaged in traditional non-paid job, i.e., housewives. Land ownership is markedly higher among Bhutia families

Table No. 1
Distribution of Mothers with Socio-demographic Variables

Variables	Rai (n= 118)		Bhutia (n=42)	
	Frequency	%	*Frequency*	%
Literacy status				
Illiterate	52	44.07	14	33.33
Literate	66	55.93	28	66.67
Level of education				
Illiterate	52	44.07	14	33.33
Class 1-4	20	16.95	7	16.67
Class 5-9	36	30.51	13	30.95
H.S.& H.S.+	10	8.47	8	19.05
Occupation				
Agriculture/Labourer/Other	21	17.80	5	11.90
Housewife	71	60.17	28	66.67
Service/Business/Student	26	22.03	9	21.43
Land ownership				
Yes	25	21.19	39	92.86
No	93	78.81	3	7.14
Amount of land				
No land	93	78.81	3	7.14
<1 acre	10	8.47	18	42.86
1-3 acres	13	11.02	19	45.24
3 acres +	2	1.69	2	4.76
Monthly income of family				
=< 2000	47	39.83	17	40.48
2001- 5000	57	48.31	13	30.95
> 5000	14	11.86	12	28.57
Monthly expenditure of family				
=< 2000	41	34.75	13	30.95
2001-5000	64	54.24	21	50.00
>5001	13	11.02	8	19.05
Parity				
Primipara	10	8.47	3	7.14
Multipara	67	56.78	29	69.05
Grand multipara	41	34.75	10	23.81

(92.9%) than Rai families (21.2%). Though, among landowners only 1.7% of Rai and 4.8% of Bhutia families have more than 3 acres of land. About 90% of Rai families and about 70% of Bhutia families have monthly income and expenditure less than Rs. 5000 per month.

Table No. 2 shows information about acceptance of antenatal care and delivery care taken during the previous pregnancy. Only 39.8% Rai women and 47.6% Bhutia women had taken antenatal care during last pregnancy.

Table No. 2
Distribution of Mothers taken Antenatal Care and Delivery Care

Information	Rai (n= 118)		Bhutia (n=42)	
	Frequency	%	*Frequency*	%
Antenatal care taken				
Yes	47	39.83	20	47.62
No	71	60.17	22	52.38
Facility availed for antenatal care				
PHSC	14	29.79	6	30.00
District Hospital	32	68.09	12	60.00
Private Clinics	1	02.13	2	10.00
Taken iron/folic acid tablets during last pregnancy				
Yes	45	38.14	16	38.10
No	73	61.86	26	61.90
Taken any special diet during last pregnancy				
Yes	112	94.92	41	97.62
No	6	05.08	1	02.38
Followed food taboos				
Yes	27	22.88	5	11.90
No	91	77.12	37	88.10
Place of delivery				
Home	94	79.66	31	73.81
Government hospital	23	19.49	11	26.19
Private hospital	1	00.85	0	00.00
Type of delivery				
Normal delivery	114	96.61	35	88.10
Caesarian delivery	4	03.39	5	11.90
Skilled attendance at delivery				
Attended	27	22.88	11	26.19
Unattended	91	77.12	31	73.81

Most of them (94.9% of Rai and 97.6% Bhutia) had taken some kind of special diet during pregnancy. In case of Rai, 79.7% deliveries have been at home whereas in Bhutia, the percentage of home delivery is 73.8. Percentage of attended delivery for Rai is 22.9 whereas it is 26.2 for Bhutia.

The contingency table for antenatal care and socio-demographic variables shows significant association with literacy status on the acceptance of antenatal care (Table No. 3). In both ethnic groups, 59.3% of literates have availed of antenatal care facilities compared to 17.6% illiterate women. At the same time, the level of education also has a significant positive trend with the proportion of women gone for antenatal care. In contrast to this, parity has a negative trend with acceptance of antenatal care. 52.9% of primipara has taken antenatal care whereas it reduces to only 14.3% for grand multipara. However, ethnicity and land ownership has not shown any significant relationship with acceptance of antenatal care during pregnancy.

The results of contingency table between utilization of skilled attendance at delivery and socio-demographic variables show almost similar results as of antenatal care (Table No. 4). It is observed that the percentage of attended delivery increased with the increased level of mother's education. In case of illiterate mothers, the percentage of attended delivery is only 12.2%, whereas for the mothers with higher secondary level of education, it is 63.6%. Attended delivery showed significant association with mother's education. Association of attended delivery with mother's occupation, monthly income and expenditure of the family has not been significant. Antenatal care taken and parity has significant association with utilization of skilled attendance at delivery.

Table Nos. 5 and 6 present logistic regression models to explain the determinants of the acceptance of antenatal care and use of safe delivery care. In case of antenatal care, when data was controlled for the education, the odds ratios (OR) for mothers with primary level of education is 1.49 and for mothers with secondary and higher level of education are 14.93 and 21.11 respectively. After controlling for ethnicity, occupation and monthly expenditure of the family, changes in OR are noted but the level of education, viz., secondary and higher level of education remained significant after controlling for other socio-demographic factors.

Similar results have been reported for utilization of skilled attendance at delivery. Model-I shows that OR for mothers with primary level of education is 1.88 and for mothers with secondary and higher level of education odds ratios are 3.16 and 12.64, respectively. Changes in the OR of different educational groups have been noted after including other variables like ethnicity, mother's occupation and monthly expenditure of the family. When controlled for parity, the odds ratios for primary, secondary and higher level of education reported to be 1.33, 1.84 and 2.80, respectively.

DISCUSSION

Relationship between selected socio-economic and demographic factors and utilization of health facilities during pregnancy and at birth was investigated using multivariate statistical techniques. Studies in different developing countries suggest that health practices during and after pregnancy are largely regulated through socio-demographic factors (Shakya *et al.*, 2001; Rodolfo *et al.*, 2002; Cindoglu and Sirkeci, 2001; Griffths and Stephenson, 2001; Chandihok *et al.*, 2006).

Table No. 3
Contingency Table for Distribution of Mothers on Antenatal Care taken with Socio-demographic Variables and Delivery Related Information

Indicators	Antenatal care not taken		Antenatal care taken		X^2 (df)	p-value
	Frequency	%	*Frequency*	%		
Ethnicity						
Rai	73	61.86	45	38.15	0.398	0.533
Bhutia	23	54.76	19	45.24	(1)	
Literacy status						
Illiterate	61	82.43	13	17.57	27.154	.0001**
Literate	35	40.70	51	59.30	(1)	
Level of education						
Illiterate	61	82.43	13	17.57	51.531	.0001**
Primary	22	75.86	7	24.14	(3)	
Secondary	11	23.91	35	76.09		
H.S.+ H.S.+	2	18.18	9	81.82		
X^2_{trend} = 45.74**						
Occupation						
Agriculture/Labourer/Other	20	83.33	4	16.67	11.540	0.0031**
Housewife	68	60.18	45	39.82	(2)	
Service/Business/Student	8	34.78	15	65.22		
Monthly expenditure of family						
=< 2000	25	53.19	22	46.81	6.973	0.0306*
2001-5000	61	68.54	28	31.46	(2)	
>5000	10	41.67	14	58.33		
X^2_{trend} = 0.089						
Monthly income of family						
=< 2000	35	58.33	25	41.67	6.436	0.0400*
2001-5000	50	68.49	23	31.51	(2)	
>5000	11	40.67	16	59.26		
X^2_{trend} = 0.089						
Does your family any land						
Yes	41	60.26	27	39.71	0.010	.9480
No	55	59.78	37	40.22	(1)	
Parity groups						
Primipara	8	47.06	9	52.94	16.471	0.0003**
Multipara	50	53.19	44	46.81	(2)	
Grand multipara	42	85.71	7	14.29		
X^2_{trend} = 14.03**						
Place of delivery						
Institutional	7	20.00	28	80.00	27.771	0.0001**
Home	89	71.77	36	28.23	(1)	
Attended or unattended delivery						
Attended	7	19.44	29	80.56	29.691	0.0001**
Unattended	89	71.77	35	28.23	(1)	
Type of delivery						
Normal	95	62.91	56	37.09	7.461	0.0063**
Caesarian	1	11.11	8	88.89	(1)	

* Significant at 0.05 level, ** Significant at .01 level.

Table No. 4
Contingency Table for Distribution of Mothers on Attended or Unattended Delivery with Socio-demographic Indicators and Antenatal Care taken During Pregnancy

Indicators	Unattended		Attended		X^2 (df)	p-value
	Frequency	%	*Frequency*	%		
Ethnicity						
Rai	93	78.81	25	21.19	0.204	0.6514
Bhutia	31	73.19	11	26.19	(1)	
Literacy status						
Illiterate	65	87.84	9	12.16	7.371	0.0066**
Literate	59	68.60	27	20.69	(1)	
Level of education						
Illiterate	65	87.84	9	12.16	16.92	0.0007**
Primary	23	79.31	6	20.69	6	
Secondary	32	69.57	14	30.43	(3)	
H.S. & H.S.+	4	36.36	7	63.64		
X^2_{trend} = 14.50**						
Occupation						
Agriculture/Labourer/Other	21	87.50	3	12.50	3.375	0.1849
Housewife	88	12.50	25	22.12	(2)	
Service/Business/Student	15	65.22	8	34.78		
Monthly expenditure of family						
=< 2000	35	74.47	12	25.53	0.598	0.742
2001-5000	71	79.78	18	20.22	(2)	
>5000	18	75.00	6	25.00		
X^2_{trend} = 0.58						
Monthly income of family						
=< 2000	45	75.00	15	25.00	1.924	0.3822
2001-5000	60	82.19	13	17.81	(2)	
>5000	19	70.37	8	29.63		
X^2_{trend} = 0.013						
Family owns land						
Yes	51	75.00	17	25.00	0.211	0.6458
No	73	79.35	19	20.65	(1)	
Taken prenatal care						
Yes	35	54.69	29	45.31	29.69	0.0001**
No	89	92.71	7	7.29	1	
					(1)	
Parity group						
Primipara	8	64.71	3	35.29	9.243	0.0098**
Multipara	63	71.28	24	28.72	(2)	
Grand multipara	42	91.84	1	8.16		
X^2_{trend} = 8.340**						

* Significant at 0.05 level, ** Significant at .01 level.

Table No. 5
Logistic Regression Results for Likelihood of Acceptance of Antenatal Care and some Important Socio-demographic Variables

Variables	Model 1		Model 2		Model 3		Model 4		Model 5	
	OR	95% CI	OR	95% CI	OR	95% CI	OR	95% CI	OR	95% CI
Level of education										
Illiterate	1.00	-	1.00	-	1.00	-	1.00	-	1.00	-
Primary	1.49	[0.53, 4.22]	1.51	[0.53, 4.26]	1.48	[0.52, 4.26]	1.25	[0.42, 3.68]	.77	[0.22, 2.71]
Secondary	14.93**	[6.05, 36.87]	15.23**	[6.12, 37.90]	13.30**	[5.19, 34.05]	12.43**	[4.80, 32.17]	7.89**	[2.74, 22,72]
H.S. & H. S.+	21.11**	[4.08, 109.38]	23.14**	[4.21, 127.15]	20.92**	[3.60, 121.49]	18.70**	[3.18, 110.01]	8.95*	[1.13, 71.08]
Ethnicity										
Rai			1.00		1.00	-	1.00	-	1.00	-
Bhutia			0.82	[0.33, 2.03]	0.80	[0.32, 2.01]	0.78	[.30, 2.05]	0.66	[0.22, 2.04]
Occupation										
Agriculture/Labourer/ Others					1.00	-	1.00	-	1.00	-
Housewife					1.92	[0.54, 6.87]	2.47	[.64, 9.51]	1.80	[0.22, 71.08]
Service/Business/ Student					2.28	[0.45, 11.49]	2.94	[54, 15.94]	1.80	[0.44, 71.08]
Monthly expenditure of family										
=< 2000							1.00	-	1.00	-
2001-5000							0.40	[0.16, 1.01]	0.43	[0.16, 1.16]
> 5000							0.77	[0.21, 2.77]	1.02	[0.23, 4.51]
Parity										
Primipara									1.00	-
Multipara									1.59	[0.33, 7.7]
Grand multipara									0.33	[0.05, 2.21]

* Significant at 0.05 level, ** Significant at .01 level.

Table No. 6
Logistic Regression Results for Likelihood of Type of Delivery (attended & unattended) and some Important Socio-demographic Variables

Variables	Model 1		Model 2		Model 3		Model 4		Model 5	
	OR	95% CI	OR	95% CI	OR	95% CI	OR	95% CI	OR	95% CI
Level of education										
Illiterate	1.00	-	1.00	-	1.00	-	1.00	-	1.00	-
Primary	1.88	[0.60, 5.87]	1.90	[0.61, 5.91]	1.87	[0.59, 5.89]	1.75	[0.55, 5.62]	1.33	[0.35, 4.96]
Secondary	3.16*	[1.24, 8.07]	3.20*	[1.25, 8.20]	2.90*	[1.09, 7.73]	2.83*	[1.05, 7.63]	1.84	[0.60, 5.64]
H.S. & H. S.+	12.64**	[3.08, 51.90]	13.50**	[3.10, 59.06]	12.51**	[2.71, 57.81]	13.80**	[2.86, 66.67]	2.80	[0.40, 19.80]
Ethnicity										
Rai			1.00	-	1.00	-	1.00	-	1.00	-
Bhutia			0.86	[0.34, 2.17]	0.85	[0.34, 2.14]	0.72	[0.29, 1.80]	1.26	[0.41, 3.85]
Occupation										
Agriculture/Labourer/ Others					1.00	-	1.00	-	1.00	-
Housewife					1.57	[0.40, 6.16]	1.80	[0.44, 7.45]	1.22	[0.28, 5.39]
Service/Business/ Student					1.72	[033, 9.03]	2.14	[0.38, 12.04]	1.60	[0.26, 9.83]
Monthly expenditure of family										
=< 2000							1.00	-	1.00	-
2001-5000							0.72	[0.29, 1.80]	.68	[0.25, 1.85]
> 5000							0.49	[0.13, 1.90]	.38	[0.07, 2.08]
Parity										
Primipara									1.00	-
Multipara									1.32	[0.30, 5.93]
Grand multipara									0.12	[0.01, 1.49]

* Significant at 0.05 level, ** Significant at .01 level.

In Skkim, except NFHS-2 (1998-1999), studies about pregnancy and delivery care are scarce. In India practices related to pregnancy and childbirth are highly influenced by prevailing societal norm, which increases the importance of community level studies. Though ethnicity has not shown significant association with pregnancy or delivery care in the present study, it has been reported to be a significant determinant of obstetric health practices in other studies (Mukhopadhyay *et al.*, 2002).

Of literate women, 59.3% have received antenatal care, which is lower than the results of NFHS-2 (about 80%). However, level of education appeared to be a significant predictor of health care utilization during pregnancy as well as at the time of delivery. Mothers with secondary level of education are about fifteen times more likely to take antenatal care than illiterate mothers (OR=14.92). In case of utilizing skilled attendance at birth, mothers with secondary level of education are about three times more likely to receive care facilities than illiterate mothers (OR=3.16). Results of the present study suggest that secondary level of education is important to determine antenatal care seeking practices, which are similar to the findings of NFHS-2. In the same way, higher secondary level of education is crucial in utilization of skilled attendance at birth (similar to the findings of NFHS-2).

Birth order has been another important obstetric factor, which significantly affects the use of safe delivery services (Obermeyer *et al.*, 1991; Dujardin,1995; Bloom *et al.*, 1999). It can be noted that the grand multipara women are less likely to avail of skilled attendance at delivery (OR=0.12). It is interesting to note that when controlled for other variables, one by one, changes have occurred in OR of level of education, but after controlling for parity considerable changes have been noted in OR of education. Result of Chi-square shows that parity has a significant association with antenatal care acceptance and utilization of skilled attendance at delivery. It is distressing that likelihood of accepting antenatal care and utilization of delivery care decreases among grand multipara whereas many studies have reported that grand multipara and primipara are at greater risk of maternal mortality and morbidity as well as having poor birth outcome (Tessema *et al.*, 2002; Bhardwaji *et al.*, 1991; Nigussie *et al.*, 2004).

A majority of women who went for antenatal check-ups utilized skilled attendance at delivery (80.56%) and most of them have opted for institutional delivery (80.00%) (Table-3). In order to ensure institutional delivery, it may therefore be advisable to promote antenatal check-ups and other necessary counselling at the grassroot level of the society. In traditional society, home delivery is preferred because of better family support, familiarity, tradition and belief, since birth is considered as a natural phenomenon which does not need institutional delivery (Griffths and Stephenson, 2001; Chandhiok, 2006; Bloom *et al.*, 1999). Awareness about the benefits of antenatal care and delivery care at community level is required, which also involves the role of traditional birth attendance in conducting home delivery on one hand and easy access to referral care, on the other.

Socio-economic factors like mother's occupation and monthly expenditure have been found to be good predictors of antenatal care and delivery care. Mothers, employed in service sectors, are 3.28 times more likely to accept antenatal care and 1.72 times more likely to avail skilled attendance at birth, in comparison to those who are agricultural labourers or other labourers. It can be argued that factors, which determine a woman's own status such as her level of education and her occupation, are decisive in health-related practices during pregnancy. It may therefore be suggested that a woman's decision-making capability about her own health is crucial for prenatal care practices.

Present cross-sectional retrospective study can be considered important as it deals with the small-scale community level data, which is again scarce. Results may suffer from some bias, which can be attributed to the small population size. There is definite need for community-based studies with large sample size to explore the role of other important factors such as women's knowledge about available health facilities, her cognitive disposition, access to health facilities, quality and cost of the service and so on.

From the present study, it can be safely argued that education directly affects women's decision about their own health, and largely determines birth outcome. The provision of health facilities do not ensure that women use them; they have to first perceive those as beneficial to them and for their unborn child. It would therefore be worthwhile to train more women in safe delivery practices, especially older women, who are well accepted in community. Interaction of health professionals with community is also needed to be increased, so that women may become more comfortable with the professional health care through progressive exposure during pregnancy. All these may contribute to an increased utilization of existing health care, thereby ensuring safe motherhood.

ACKNOWLEDGEMENTS

The authors wish to thank the study participants (both from the Rai and the Bhutia communities) in the villages of Ranka, Sikkim for their active participation. Thanks are due to Mr. Sanjeev Bakshi of Indian Statistical Institute, Kolkata for his positive contribution in statistical analysis.

BIBLIOGRAPHY

AbouZahr, C. and T. Wardlaw. 2004. 'Maternal mortality in 2000: Estimates developed by WHO, UNICEF and UNFPA'. Geneva: World Health Organization.

Agresti, A. 1990. *Categorical Data Analysis*. New York: John Wiley and Sons.

Altman, D.G. 1991. *Practical Statistics for Medical Research*. Madras: Chapman and Hall.

Bhardawji, N., S.B. Hasan, M. Yunus and M. Zahara. 1991. 'High risk pregnancy and relation with maternal care receptivity: a rural study from India'. *Journal of Royal Society of Health* 111: 43-46.

Bhatia, J.C. and J. Cleland. 1997. 'Determinants of maternal care in a region of south India'. *Health Transition Review* 5: 127-142.

Bloom, S.S., T. Lippeveld and D.Wypij. 1999. 'Does antenatal care make a difference to safe delivery? Study in urban Uttar Pradesh, India'. *Health Policy and Planning* 14: 38-48.

Bolam, A., D.S. Manandhar, P. Shrestha, B. Manandhar, M. Ellis and A.M. Costello. 1997. 'Maternal mortality care utilization in Kathmandu valley'. *The Journal of Nepal Medical Association* 35: 122-129.

Carlaugh, M. 1997. 'More than hospitals are needed in Nepal'. *Safe Motherhood* 24: 9.

Chandhiok, N., B.S. Dhillon, I. Kambo and N. Saxena. 2006. 'Determinants of antenatal care in rural areas of India: a cross-sectional study from 28 districts (a ICMR task force study)'. *The Journal of Obstetric and Gynecology of India* 56: 47-52.

Chandrashekar, S., R.S. Rao, N.S. Nair and P.R. Kutty. 1998. 'Sociodemographic determinants of antenatal care'. *Tropical Doctors* 28: 206-209.

Cindoglu, D. and I. Sirkeci. 2001. 'Variables that explain variation in prenatal care in Turkey; social class education and ethnicity visited'. *Journal of Biosocial Science* 33: 261-270.

Coyaji, B.J. 1991. 'Maternal mortality and morbidity in the developing countries like India'. *Indian Journal of Maternal and Child Health* 2: 3-9.

Dhamala, R.R., C.D. Rai, M.S. Dutta and S. Ghatak (eds). 1993. *People of India: Sikkim*. Vol. 39. Calcutta: Anthropological Survey of India and Seagull Books.

Dujardin, B., G. Clarysee, B. Criel, V. De Brouwere and N. Waganta. 1995. 'The strategy of risk approach in antenatal care: evaluation of the referral compliance'. *Social Science and Medicine* 40: 529-535.

Elo, I.T. 1992. 'Utilization of maternal health-care services in Peru: The role women's education'. *Health Transition Review* 2: 49-69.

Griffiths, P. and R. Stephenson. 2001. 'Understanding users perspective of barriers to maternal health care use in Maharashtra, India'. *Journal of Biosocial Science* 33: 339-359.

Jejeebhoy, S.J. and S.R. Rao. 1995. 'Unsafe motherhood: A review of reproductive health', in M. Dasgupta, C.L. Chen and T.N. Krishnan (eds.), *Women's Health in India: Risk and Vulnerability*. Bombay: Oxford University Press.

Lebiratos, P., B.G. Link and J.L. Kese. 1988. 'The measurement of social class in epidemiology'. *Epidemiology Review* 10: 87-121.

Matthews, Z., S. Mahendra, A. Kilaru and S. Ganapathy. 2001. 'Antenatal care, care seeking and morbidity in rural Karnataka, India: results of a prospective study'. *Asia Pacific Population Journal* 16: 11-28.

Ministry of Health and Family Welfare (MOHFW). 1998. *Manual on Community Needs Assessment Approach in Family Welfare Programme*. New Delhi: Department of Family Welfare, MOHFW.

Ministry of Health and Family Welfare (MOHFW). 1997. *Reproductive and Child Health Programme: Schemes for Implementation*. New Delhi: Department of Family Welfare, MOHFW.

Mukhopadhyay, S., S. Ray, S. Ghosh and J.C. Bhatia. 2002. 'Obstetric morbidity and socio-demographic factors in rural West Bengal, India'. *The European Journal of Contraception and Reproductive Health Care* 7: 41-52.

National Family Health Survey (NFHS-2) India. 1998-99 Sikkim. 2001. 'Indian Institute of Population Sciences, Mumbai'. Measure DHS+, ORC Macro. USA.

Nigussie, M., D.H. Mariam and G. Mitike. 2004. 'Assessment of safe delivery service utilization among women of childbearing age in north Gondar Zone, northwest Ethiopia'. *Ethiopian Journal of Health Development* 18: 145-152.

Obermeyer, C.M. and J.E. Potter. 1991. 'Maternal health care utilization in Jordan: A study pattern and determinants'. *Studies in Family Planning* 22: 177-187.

Palaniappan, B. 1995. 'Role of antenatal care in safe motherhood'. *Journal of Indian Medical Association* 93: 53-54.

Pebley, A.R., N. Goldman and G. Rodriguez. 1996. 'Prenatal and delivery care and childhood immunization in Guatemala: do family and community matter?' *Demography* 33: 231-247.

Petrou, S., E. Kupek., S. Vause and M. Maresh. 2001. 'Clinical provider and sociodemographic determinants of the number of antenatal visits in England and Wales'. *Social Science and Medicine* 52: 1123-1134.

Rodolfo, A.B. and J.A. Ross. 2002. 'Rating maternal and neonatal health services in developing countries'. *Bulletin of WHO* 80 (9).

Shakya, K. and C. McMurry. 2001. 'Neonatal mortality and maternal health care in Nepal: searching for patterns for patterns of association'. *Journal of Biosocial Science* 33: 87-105.

Tessema, F., M. Asefa and B. Ayele. 2002. 'Mother's health service utilization and health care seeking behaviour during infant rearing: a longitudinal community based study, southwest Ethiopia'. *Ethiopian Journal of Health Development* 16 (special issue): 51-58.

Wagle, R.R., S. Sabroe and B.B. Nielsen. 2004. 'Socioeconomic and physical distance to the maternity hospital as predictors for place of delivery: An observation from Nepal'. *Bio Med Central Pregnancy and Child Birth* 4.

World Health Organization and United Nations Children's Fund. 1996. 'Revised 1990 estimates of maternal mortality: a new approach by WHO and UNICEF'. Geneva: WHO.

World Health Organization. 1994. 'Mother-baby package: implementing safe motherhood in countries'. Geneva: World Health Organization.

19

SEASONAL VARIATION IN BLOOD PRESSURES AND HYPERTENSION AMONG THE BHUTIAS OF SIKKIM

Sobhanjan Sarkar and Barun Mukhopadhyay

INTRODUCTION

In studying diseases and disorders among human populations, both epidemiological and medical anthropological approaches have been adopted. The chief aim of these approaches has been to unravel the etiology of the diseases and to identify the nature and extent of inter-relationships among its bio-cultural including behavioural and environmental determinants. Non-communicable diseases (NCDs) accounted for almost 60% of the global deaths and 46% of the global burden of disease in 2000 (WHO, 2002a). CVDs are one of the major categories of NCDs. It is estimated that CVDs will be the most chronic factor as a cause of disability-adjusted life years (DALYs) lost in 2020 (WHO, 2002a). World Health Organization (WHO) declared CVDs as the most serious epidemic of the world after its sharp increase between 1940 and 1967 (WHO, 1993).

The investigators of the Framingham Heart Study (FHS) first used the term 'risk factor' in case of heart diseases, a concept that is now widely accepted and used (Kannel *et al.*, 1961). Mackay and Mensah (2004) mention that over 300 risk factors are found to be associated with CHD and stroke, of which high blood pressure or hypertension is the most important modifiable biological risk factors. The WHO estimates that globally 600 million people with high blood pressure are at risk of heart attack, stroke and cardiac failure (WHO, 2002b).

Lifestyle is widely accepted as being one of the four factors which contribute to the health status of an individual, the other three being heredity, environment and health care (Boyd and Chaseling, 1992). The relationship between blood pressure/hypertension and lifestyle has been an established phenomenon. Many authors have studied the relationship between hypertension and socio-economic status, psycho-social stress, smoking, alcoholism, dietary habit, physical activity pattern and so on among different populations. Again, the association of hereditary factors, i.e., genetic factors, with hypertension has also been examined in different populations. However, the inter-relationship between climatic conditions and blood pressure/hypertension received relatively little attention globally. Seasonal influence on blood pressure was first described in Britain (Rose, 1961), where mean blood pressure value is found to be higher in spring than winter. On the contrary, Heller *et al.* (1978) showed higher mean DBP value in the winter season among the UK population. Similarly, among the Japanese, winter blood pressures are reported to be higher than summer (Hata *et al.*, 1982). Several other studies have shown higher mean value of blood pressures and prevalence of hypertension among adults, young adults and children in the winter season (Kiefe *et al.*, 1997; Shaper *et al.*, 1981; Jenner *et al.*, 1987; Prineas *et al.*, 1980). Seasonal variation was reported to be the most important influencing factor in the Copenhagen City Heart Study (Andersen *et al.*, 2002). It has been depicted in a study by Charach *et al.* (2004) that cold weather played significant role in raising blood pressure among the patients of hypertension in Israel. Ghosh *et al.* (2003) also suggested that normal individuals show higher blood pressure during the winter season. In another study, Isezuo (2003) reported higher hospitalization in winter season due to hypertension than summer among the Nigerians. Very recently some other studies showed increase in blood pressure in winter season than summer (Cheng *et al.*, 2006; Mori *et al.*, 2006).

In India, many studies have been undertaken to examine the association of blood pressure and hypertension with socio-economic status, behavioural factors, diet, physical activity, psycho-social stress, but study on seasonal variation of blood pressure is extremely rare. A patient-control study in Chandigarh showed higher blood pressure in winter season for the patients (Sharma *et al.*, 1990). Seasonal variation in blood pressure has also been reported among a cross-section of a North Indian population (Narang and Wasir, 1996).

While several hospital-based studies have been done regarding the summer-winter variation in blood pressure, population-based studies are found to be comparatively less in global perspectives in general and India in particular. Literature review suggests scarcity of information on seasonal variation in blood pressure profile in any Indian population. Such study has never possibly been conducted among any tribal population in India. Hence, the present study has been undertaken to examine the seasonal variation, and its probable reasons, in blood pressure and hypertension profile among the Bhutias, a tribal community living in the Himalayan state of Sikkim in North-East India.

MATERIALS AND METHODS

For the present study, the Bhutias, residing in the town of Gangtok and its neighbourhood, and those residing in Ralong, Namlung and Lingding Revenue Blocks of South District were selected as urban and rural study participants, respectively. The Bhutias have Mongoloid physical features. They have a language of their own, which is known as *Sikkimese*. However, they also speak in Nepali. Rural Bhutias are mostly agriculturists while the urban Bhutias are predominantly engaged in service and business. After the merger of Sikkim with the Indian Union in 1975, Bhutias are increasingly taking advantage of expanding developmental programmes in the state. The Sikkimese Bhutias, especially the study participants, are exclusively Buddhists. Monogamy is the general norm among them, but some cases of polygamous marriages are present. Polyandry and consanguineous marriages are socially not acceptable among them.

The Bhutias are mostly non-vegetarians and thrive on carbohydrates, red meat and milk products. Beef (*langsha*) and pork (*faksha*) are consumed in large quantities. They prepare a traditional item from milk of buffalo named *churpi*, which is very rich in saturated fat. Irrespective of sex, consumption of alcohol is very much common among the Bhutias. They drink a variety of alcoholic beverages, both indigenous ones and those liquors available in the market. The indigenously made millet beer *chang* and hard drink *Raxi* are popular among them. Further, they consume tea added with salt, but as a result of urbanization, the use of sugar is now becoming common among the urban Bhutias.

Bhutias of both sexes aged 20 years and above were considered for the study. A total of 200 adults (100 male and 100 female) from 100 randomly selected families and 228 adults (103 male and 95 female) from 80 families constituted the urban and rural study participants, respectively. In case of rural area, all the families in selected villages were considered. One adult male and one adult female, preferably the household head and his wife, participated from each family. In the absence of such a combination, other combinations such as mother-son, father-daughter, and the like were also considered. In the rural area, to increase the sample size, more than one male/female from a family has been selected. Correct ages of the study participants, particularly those older ones, for whom accurate age record was absent, were estimated with the help of a traditional Bhutia calendar.

Socio-economic, anthropometric and blood pressure data were collected in winter (January-February) as well as summer (May-June) seasons. Data on socio-economic variables such as education, occupation and so on were collected with the help of standard questionnaires by interviewing each study subject. Anthropometric measurements considered for the study included stature and weight and were measured following standard techniques (Lohman *et al.*, 1988). One of the authors (SS) recorded all the measurements for the males and females in each season. The body mass index (BMI) was calculated as weight/stature^2 (kg/m^2).

In each season, blood pressure (BP) measurements of each study participant were recorded for two consecutive times following standard methods, i.e., systolic blood pressure (Korotkoff phase I) and diastolic blood pressure (Korotkoff phase V). A stethoscope and a mercurial sphygmomanometer were used. The blood pressures were taken on the left upper arm with the subject in sitting position after a 10 minutes rest. Blood pressure readings (both SBP and DBP) were taken twice and the average is considered for analyses. Written informed consent to participate in the study was obtained from each study participant. Hypertension has been defined following Joint National Committee on Prevention Detection Evaluation and Treatment of High Blood Pressure, National High Blood Pressure Education Program Coordinating Committee (JNC-7) criteria (SBP = 140 mm Hg and/or DBP = 90 mm Hg) (Chobanion *et al.*, 2003).

All statistical analyses have been done with the help of SPSS 11.0 for windows. Before the inferential statistical analyses, log transformation was done for the blood pressure values to reduce the skewness and kurtosis. The effect of age was removed from SBP and DBP through linear regression analysis and the standardized residuals obtained in the regression analysis were used in subsequent inferential analysis. Mean and standard deviations of the SBP and DBP were calculated and the significance of difference between seasons with the help of t-test. Seasonal difference in prevalence of hypertension was evaluated through test of equality of proportions. Stepwise multiple regression analysis was performed to identify the body fat related measure(s) that significantly predicts SBP and DBP. Logistic regression analysis was carried out to estimate the likelihood of hypertension in winter season compared to that in summer.

RESULTS

Descriptive statistics (mean and standard deviation) of SBP and DBP are shown in Table No. 1 according to season, habitat and sex. It has uniformly been found that the mean SBP and DBP are higher in the winter season, irrespective of habitat and sex, except for rural males. While the mean SBP value is found to be significantly higher among the urban females in the winter season ($p<0.05$), mean DBP value is found to be significantly higher in case of urban study participants ($p<0.05$), irrespective of sex.

Table No. 1
Descriptive Statistics (mean ± sd.) of Blood Pressure Variables (in mm Hg), by Sex, Habitat and Season

Blood Pressure	Season	Urban		Rural	
		Male	Female	Male	Female
Systolic	Winter	134.56 ± 19.56	130.60 ± 18.53*	129.40 ± 10.94	131.46 ± 16.30
	Summer	134.47 ± 19.49	122.33 ± 14.45	130.53 ± 11.77	131.12 ± 16.71
Diastolic	Winter	88.21 ± 10.63 *	85.30 ± 10.86 *	81.42 ± 6.19	81.86 ± 6.98
	Summer	85.58 ± 7.95	81.79 ± 9.11	82.01 ± 5.82	81.49 ± 6.79

* Significantly higher in winter at < 0.05 level

Table No. 2 depicts the prevalence of hypertension categories according to JNC-7 criteria. It has generally been found irrespective of sex and season that urban study participants show higher prevalence of hypertension than their rural counterparts. Irrespective of sex, urban study participants show significantly higher prevalence of hypertension in winter season ($p< 0.05$) than that in summer.

Stepwise multiple regression analysis was performed to identify the body fat related factors that significantly predict age adjusted log-transformed SBP (Table 3). Of the two independent variables, i.e., weight and BMI, the former is found to be a significant predictor in case of males

Table No. 2
Prevalence of Hypertension (%) among the Study Population, by Sex, Habitat and Season

Hypertension Category	Urban				Rural			
	Male		Female		Male		Female	
	Winter	Summer	Winter	Summer	Winter	Summer	Winter	Summer
Normal	12.0	8.1	23.0	29.5	13.6	12.9	23.2	19.3
Pre-hypertensive	43.0	59.3 #	43.0	47.4	76.7	72.0	49.5	59.0
Hypertensive	45.0 *	32.6	34.0 *	23.1	9.7	15.1	27.4	21.7

* Significantly higher in winter at < 0.05 level
\# Significantly higher in summer at < 0.05 level

(b = 0.249, $p < 0.05$) and females (b = 0.313, $p < 0.01$) of urban study area during the winter season. In summer, while weight significantly predicts SBP among the urban as well as rural males (b = 0.299 and 0.346, respectively $p < 0.01$), no significant predictor has been identified in case of females in summer, irrespective of habitat.

Table No. 3
Stepwise Multiple Regression Analysis using Body Fat Related Variables as Independent and Log-transformed, Age-adjusted SBP as Dependent Variables

Season	Habitat	Sex	Significant Predictor	β	p	R^2
Winter	Urban	Male	Weight	0.249	0.013	0.062
	Rural	Female	Weight	0.313	0.002	0.098
	Urban	Male	-	-	-	-
	Rural	Female	BMI	0.213	0.038	0.045
Summer	Urban	Male	Weight	0.299	0.005	0.089
	Rural	Female	-	-	-	-
	Urban	Male	Weight	0.346	0.001	0.119
	Rural	Female	-	-	-	-

Table No. 4 shows the results of stepwise multiple regression analysis using log-transformed and age adjusted DBP as dependent, and body fat related variables as independent variable. Weight is found to the only significant predictor for DBP irrespective of sex and habitat, except rural females, in both seasons.

Further, the SBP and DBP values were adjusted for significant body fat related measures and habitat. T-tests were performed to find the significance of difference in mean values between winter and summer and the results has been shown in Table No. 5. While mean difference is found to be significantly higher in winter season in case of females for both SBP ($t = 2.858$, $p < 0.01$) and DBP ($t = 2.081$, $p < 0.05$), winter-summer difference is case of males is found to be non-significant.

Table No. 6 depicts the result of logistic regression analysis for estimating the likelihood of hypertension in winter season in comparison with summer, by sex and habitat. For urban study participants, both males (OR = 1.695, 95% CI = 0.931 – 3.084) and females (OR = 1.472, 95% CI = 0.771 – 2.810) have significantly higher chance of hypertension. However, in case of rural study participants, while females show significantly higher

Table No. 4
Stepwise Multiple Regression Analysis using Body Fat Related Variables as Independent and Log-transformed, Age-adjusted DBP as Dependent Variables

Season	Habitat	Sex	Significant Predictor	β	p	R^2
Winter	Urban	Male	Weight	0.265	0.008	0.070
	Rural	Female	Weight	0.221	0.027	0.049
	Urban	Male	Weight	0.275	0.005	0.076
	Rural	Female	-	-	-	-
Summer	Urban	Male	Weight	0.301	0.005	0.091
	Rural	Female	Weight	0.232	0.038	0.054
	Urban	Male	Weight	0.384	0.000	0.148
	Rural	Female	-	-	-	-

Table No. 5
Test of Significance of Difference (t-test) of Mean Blood Pressure Values between Two Seasons after Adjusting for Habitat and Significant Body Fat Related Predictors, by Sex

Sex	Blood Pressure	t	df.	p
Male	SBP	0.862	379	0.389
	DBP	0.436	379	0.663
Female	SBP	2.858	356	0.005
	DBP	2.081	356	0.038

Table No. 6
Result of Logistic Regression Analysis for Likelihood of Hypertension in Winter Season in Comparison with Summer, by Habitat and Sex

Season	Urban				Rural			
	Male		Female		Male		Female	
	OR	CI (95%)	OR	CI (95%)	OR	CI (95%)	OR	CI (95%)
Summer	Ref.		Ref.		Ref.		Ref.	
Winter	1.695	0.931-3.084	1.472	0.771-2.810	0.607	0.255-1.441	1.361	0.683-2.713

likelihood (OR = 1.361, 95% CI = 0.683 – 2.713) of hypertension, males show higher value but the likelihood is found to be non-significant (OR = 0.607, 95% CI = 0.255 – 1.441).

DISCUSSION

Due to its location at the foothill of the Himalayas, the state of Sikkim remains relatively cooler throughout the year. Nevertheless, winter-summer difference in temperature can distinctly be observed in many parts of the State. The mean temperature in the winter season ranges from 1° C to 9.5° C (recorded in January) and that of summer is – 28° C (recorded in June). The annual rainfall also remains relatively higher in the State of Sikkim (between 2000 mm and 4000 mm) (GIS, 2005). The winter-summer distinctiveness can also be noticed from few lifestyle-related variables such as food habit (wider variety of food items are available in the winter season), behavioural pattern (smokers/alcohol consumers tend to increase number of cigarettes/amount of alcohol per day), daily activity level (generally found to be lower in winter) of the individuals and so on. Effect of those lifestyle-related variables on communicable as well as non-communicable diseases and its risk factors is well established, though the investigation on seasonal variation in any of the disease related profiles has never been attempted in any of the communities residing in Sikkim.

The present study clearly demonstrates the seasonal variation in blood pressure as well as hypertension profile of the Bhutias of Sikkim. The results show significantly higher mean value of both SBP and DBP in winter season irrespective of sex and area of habitat which corroborates the findings by Silva *et al.* (2006) among the Caboclo population of Brazil. Ghosh *et al.* (2003) also reported increase in SBP and DBP during the cold season among the Nepalese. Again, in the present study, weight is found to be a significant predictor of both SBP and DBP, irrespective of season, area of residence and sex. Bhutias who are residing in the rural area are mostly agriculturists and are used to spend major part of the day in agricultural field during the summer. Eventually, the physical activity level remains relatively higher and consequently they loose more energy. In addition, the rate of sodium excretion remains higher during summer months. Moreover, the availability of food materials remains less in variety in the market. Bhutias living in the urban area also experience higher excretion of sodium in the summer compared to winter. In the winter season, a wide variety of food items are available in the market which they consume frequently. Red meat, especially beef, is consumed in a higher amount to keep their body warm. Additionally, smokers and alcohol drinkers increase the frequency of smoking and alcohol consumption in winter. Bhutias are used to drink tea added with salt. Though it has been observed in the urban areas that many people have started using sugar in place of salt in tea, majority of them are still consuming tea with salt. In the rural area, as well, Bhutias are still maintaining their traditional habit of consuming tea with salt. Irrespective of area of residence, i.e., urban or rural, and sex, the study participants have reported higher consumption of tea with salt in the winter season than that in summer. All of these lifestyle-related modifications influence the weight gain as well as increase in the sodium concentration in blood in winter season and, as a result, blood pressure increases. Similar conclusions have been drawn by Brennan *et al.* (1982) among the UK population and Argiles *et al.* (2004) among the French people. Our study clearly shows effect of weight on blood pressure profiles, which corroborates with the study done by Hancock *et al.* (1991). The present study demonstrates significant seasonal difference in blood pressure profile even after adjusting for habitat and body fat related measures, i.e., weight and BMI, only among the females.

It is to be noted that action of some of the enzymes/proteins, like catecholamines may play an important role in regulating the blood pressure in different seasons as demonstrated by Ghosh *et al.* (2003). Rosenthal (2004) also pointed out that the lower blood pressure in warm temperatures is attributed to cutaneous vasodilatation and loss of water and salt from sweating. Such possibilities, whether existing in the study community, will be interesting to examine. In the same token, large-scale studies would be of great importance in examining the specific cause(s) of adverse profile of blood pressure and other CVD risk factors, in relation to different seasons, among the Bhutias, as well as other communities of Sikkim.

ACKNOWLEDGEMENTS

The authors are indebted to the study participants for their un-hesitated co-operation in the study. The Indian Statistical Institute is thankfully acknowledged for financial support.

BIBLIOGRAPHY

Andersen, U.O., J.H. Henriksen, G. Jensen and Copenhagen City Heart Study Group. 2002. 'Sources of measurement variation in blood pressure in large-scale epidemiological surveys with follow-up'. *Blood Pressure* 11: 357-365.

Argiles, A., G. Mourad and C. Mion. 1998. 'Seasonal changes in blood pressure in patients with end-stage renal disease treated with hemodialysis'. *The New England Journal of Medicine* 339: 1364-1370.

Boyd, R. and J. Chaseling. 1992. 'Adolescent student blood pressure change upon transmission from secondary to tertiary education'. Paper presented at Science Health and Environmental Education Research – A symposium in the Australian Association for Research in Education of New Zealand Association for Research in Education Conference. Victoria, Australia: Deakin University.

Brennan, P.J., G. Greenberg, W.E. Miall and S.G.Thompson. 1982. 'Seasonal variation in arterial blood pressure'. *British Medical Journal* 285: 919-923.

Charach, G., P.D. Rabinovich and M. Weintraub. 2004. 'Seasonal changes in blood pressure and frequency of related complications in elderly Israeli patients with essential hypertension'. *Gerontology* 50: 315-321.

Cheng, L.T., H.Y. Jiang, L.J. Tang and T. Wang. 2006. 'Seasonal variation in blood pressure of patients on continuous ambulatory peritoneal dialysis'. *Blood Purification* 24: 499-507.

Ghosh, A., T. Pramanik and P. Roychowdhury. 2003. 'Seasonal variation of blood pressure in young normotensives'. *Nepal Medical College Journal* 5: 100-101.

GIS (Geographical Information of Sikkim). 2005. 'Geographical Information of Sikkim'. Home Department. Government of Sikkim. http://sikkim.nic.in/homedept/geog.htm (accessed on 02.01.2007).

Hancock, R.L., F.A. Tylavasky, R. Moore and J.J. Anderson. 1991. 'Heredity and environmental influences on blood pressure values of premenopausal women and their college-age daughters'. *Journal of American College of Nutrition* 10: 376-382.

Hata, T., T. Ogihara, A. Maruyama, H. Mikami, M. Nakamura, T. Naka, Y. Kumahara and C.A. Nugent. 1982. 'The seasonal variation of blood pressure in patients with essential hypertension'. *Clin. Exp. Hyperten. A* 4: 341-354.

Heller, R.F., G. Rose, H.D. Tunstall-Pedoe and D.G.S. Christie. 1978. 'Blood pressure measurement in the United Kingdom Heart Disease Prevention Project'. *Journal of Epidemiology and Community Health* 32: 235-238.

Isezuo, S.A. 2003. 'Seasonal variation in hospitalisation for hypertension-related morbidities in Sokoto, north-western Nigeria'. *International Journal of Circumpolar Health* 62: 397-409.

Jenner, D.A., D.R. English, R. Vandongen, L.J. Beilin, B.K. Armstrong and D. Bunbar. 1987. 'Environmental temperature and blood pressure in 9-year-old Australian children'. *Journal of Hypertension* 5: 683-686.

Kannel, W.B., T.R. Dawber, A. Kagan, N. Revotskie and J. Stokes. 1961. 'Factros of risk in the development of coronary heart disease – six-year follow-up experience. The Framingham Study'. *Annals of Internal Medicine* 55: 33-50.

Kiefe, C.I., O.D. Williams, D.E. Bild, C.E. Lewis, J.E. Hilner and A. Oberman. 1997. 'Regional disparities in the incidence of elevated blood pressure among young adults: the CARDIA study'. *Circulation* 96: 1082-1088.

Lohman, T.G., A.F. Roche and R. Martorell. 1988. *Anthropometric Standardization Reference Manual*. Champaign, Illinois: Human Kinetics Books.

Mackay, J., G.A. Mensah (with S. Mendis and K. Greenlund). 2004. 'The Atlas of Heart Disease and Stroke'. Geneva: World Health Organization.

Mori, H., H. Ukai, H. Yamamoto, S. Saitou, K. Hirao, M. Yamauchi and S. Umemura. 2006. 'Current status of antihypertensive prescription and associated blood pressure control in Japan'. *Hypertension Research* 29: 143-151.

Narang, R. and H.S. Wasir. 1996. 'Seasonal variation in the incidence of hypertension and coronary artery disease'. *International Journal of Cardiology* 56: 90-92.

Prineas, R.J., R.F. Gillum, H. Hiribe and P.J. Hannan. 1980. 'The Minneapolis children's blood pressure study. 1: Standards of measurements for children's blood pressure'. *Hypertension* 2 (supplement I): 118-124.

Rose, G. 1961. 'Seasonal variation in blood pressure in man'. *Nature* 189: 235.

Rosenthal, T. 2004. 'Seasonal variation in blood pressure'. *American Journal of Geriatric Cardiology* 13: 267-272.

Shaper, A.G., S.J. Pocock, M. Walker, N.M. Cohen, C.J. Wale and A.G. Thomson. 1981. 'British Regional Heart Study: cardiovascular risk factors in middle-aged men in 24 towns'. *British Medical Journal* 283: 179-186.

Sharma, B.K., S. Sagar, G.K. Sood, S. Varma and O.P. Kalra. 1990. 'Seasonal variations of arterial blood pressure in normotensive and essential hypertensives'. *Indian Heart Journal* 42: 66-72.

Silva, H.P., G.D. James and D.E. Crews. 2006. 'Blood pressure, seasonal body fat, heart rate, and ecological differences in Caboclo populations of the Brazilian Amazon'. *American Journal of Human Biology* 18: 10-22.

WHO (World Health Organization). 1993. 'World Health Statistics Quarterly'. Vol. 46. Geneva: World Health Organization.

WHO (World Health Organization). 2002a. 'Health Situation in the South-East Asia Region 1998-2000'. Regional Office for South-East Asia. New Delhi: World Health Organization.

WHO (World Health Organization). 2002b. 'Cardiovascular Disease-Prevention and Control'. WHO CVD strategy, 2001/2002. Geneva: World Health Organization.

20

URBAN LIVING AND HYPERTENSION AMONG THE LEPCHAS OF SIKKIM

Barun Mukhopadhyay and Susmita Mukhopadhyay

INTRODUCTION

Human biological and epidemiological research on blood pressure and hypertension has been undertaken for long, focusing largely on cross-cultural comparisons, adaptability to certain environmental conditions and the influences on blood pressure of various cultural, behavioural and psychological changes concomitant with modernization and migration (Gerber and Halberstein, 1999). It is well documented, therefore, that the blood pressure level of a population is influenced by a variety of biological, behavioural and socio-cultural factors (Lowenstein, 1961; Hanna and Baker, 1979; McGarvey and Baker, 1979; Katz *et al.*, 1980; Rao, 1980; McGarvey and Schendel, 1986; Dressler,1999; Crews and Williams, 1999). Identification of genetic and environmental correlates of blood pressure in human population groups inhabiting varied ecological and cultural settings is considered to be helpful in gaining insights into the etiological factors for determining essential hypertension, which is a major public health problem for the developed, as well as in recent times, for developing countries such as India.

It is observed that factors that determine blood pressure levels in traditional as against modernized populations vary (Epstein and Eckhoff, 1967; Ward, 1983; Halberstein, 1999). Studies also indicated that even among the traditional populations, major factors affecting blood pressure level do not remain similar (Ward, 1983). Evidence exists that especially when the traditional populations are exposed to *in situ* modernization or migrate to relatively modern locales, they experience higher mean blood pressure or higher proportion of hypertensives (Neilson and Williams, 1978; Marmot, 1979).

Epidemiological studies of blood pressure among Indian populations inhabiting contrasting ecological zones, having varied genetic ancestry and socio-cultural characteristics are relatively rarely available. Some studies, however, have been conducted on epidemiological profile of blood pressure among socio-culturally distinguishing populations inhabiting the urban and rural areas of West Bengal plains (Mukherjee *et al.*, 1988; Majumder *et al.*, 1990, 1994). Furthermore, epidemiological profile of blood pressures among the Lepchas inhabiting a rural area of Sikkim has been studied by Mukhopadhyay *et al.* (1996). Majumder *et al.* (1994) observed much higher prevalence of hypertension in a sedentary population of Kolkata.

In view of the above, the present study evaluates the blood pressure profile in relation to some selected biological (age, sex, body dimension, disease status), behavioural (tobacco use, alcohol intake, salt intake, oral contraceptive use) and socio-economic (education) factors among the Lepchas inhabiting Gangtok, the capital town of the state of Sikkim, and its neighbourhood.

MATERIALS AND METHODS

The Lepchas, indigenous to the Sikkim Himalayas, are a tribal population with Mongoloid physical features (Gorer, 1938). They, however, call themselves *Rongkup,* which means 'sons of snowy peak' (Tamsang, 1983). The Lepchas have a language of their own which is known as *Rongring,* included in the Tibeto-Burman family of languages (Grierson, 1927). The Lepchas were predominantly hunter-gatherers and shifting hill cultivators in the past; at present a sizable majority of them are settled agriculturists. Some researchers suggested that settled agriculture began to be practised by the Lepchas of Sikkim-Darjeeling probably during mid-nineteenth century (Das, 1978). Furthermore, with changing times and the spread of education, especially since Sikkim's merger with Indian Union in 1975, Lepchas are increasingly found to be engaged also in non-agricultural occupations. In Sikkim, the Lepchas predominantly profess the Tibetan form of Lamaist Buddhism, while few have adopted Christianity. Monogamy is the marital norm for the Lepchas in general, though polygamous marriages are not rare. They, however, do not practise polyandry; consanguineous marriages are not socially acceptable and, therefore, not found to occur among them.

The Lepchas are distributed throughout the mountainous State of Sikkim. They are found to reside in both urban and rural areas of the State, though concentrated heavily in the latter. For the present investigation, Lepchas inhabiting the capital town of Gangtok and its neighbourhood was chosen as the study population.

The town of Gangtok in the east district is the largest urban centre in Sikkim. Since 1975, when it emerged as the capital of an Indian State, the magnitude of development input, in term of infrastructure, trade and commerce, housing, health and education, tourism and so on in Gangtok has enormously increased. For economic, administrative and other reasons, Gangtok has witnessed an enormous growth of population, largely due to migration from different areas of the State as well as from areas outside the State. It has become a place, which accommodated diverse ethnic communities, permanent and temporary settlers in a cosmopolitan mould. Sizable proportions of Lepchas who reside in Gangtok town are migrants either from rural areas of different districts of Sikkim or from hill subdivisions of the Darjeeling district of West Bengal. A large number of them are employed in government and other establishments, while a relatively lesser number of them are engaged in business. Lepchas, inhabiting Gangtok in substantial numbers, have adopted the urban way of life with better education, considerable change in their traditional dietary practices, increasing exposure to mass media and so on. Consumption of a variety of alcoholic beverages, including traditional millet beer (*chi*), is also very common among them. Consumption of salt with tea, in addition to its use in cooked food items freely, remains a regular feature.

Buddhist Lepchas of both genders, aged 16 years and older, inhabiting Gangtok town and its neighbourhood (at an altitudinal range of 4,000 ft. to 6,500 ft.) continuously for 10 or more years were selected for this study. Overall, 138 adults, engaged primarily in service sector, were considered for the present study. Study subjects by age groups and gender are shown in Table No. 1. Age was estimated with the help of a traditional Lepcha calendar, which follows a 12-year cycle with one year periods, especially when accurate age records did not exist.

Table No. 1
Sample Size by Age and Gender

Age Group (Years)	Male	Female	Total
≤ 29	**26**	**27**	**53**
30 – 39	24	20	44
40 – 49	13	11	24
50 – 59	6	5	11
≥ 60	**3**	**3**	**6**
Total	72	66	138

The same investigator (SM) took two consecutive blood pressure readings, both systolic and diastolic, on the left arm of each subject in the sitting position after a ten-minutes rest. A single mercury sphygmomanometer was used for measuring blood pressure from all the subjects. All blood pressure and anthropometric measurements were taken at a time within 8:00 and 14:00 hours. Data on body dimensions included the following anthropometrtic measurements: stature, weight, upper arm girth, calf girth, triceps and subscapular skinfold thicknesses. Following standard techniques (Lohman *et al.*, 1988), one male investigator (BM) took anthropometric measurements of males while one female investigator (SM) took those of females. From each subject, information on age, gender, level of education, self-reported disease status, current use of anti-hypertensive drugs, use of oral contraceptives by married non-menopausal females, smoking, tobacco chewing and alcohol intake were collected. Salt consumption pattern among the subjects was noted in terms of free use of salt as well as its use in cooked food items and with tea. Degree of salt intake was assessed in a semi-quantitative manner by noting the number of cups of tea consumed per day added with salt. Frequency distribution of subjects by gender in respect of categorical variables is shown in Table No. 2. The averages of two systolic blood pressure (SBP) and two diastolic blood pressure (DBP) readings were used for analyzing blood pressure data. The blood pressure values showed significant skewness and curtosis. Substantial reduction of skewness and curtosis values, however, was achieved by log transformation of SBP and DBP values. Therefore, logarithmically transformed values of both SBP and DBP have been used for the subsequent analyses.

In order to determine the effect of categorical secondary variables (viz., education, oral contraceptive use, disease status and gender) on the primary variables (viz., SBP and DBP), effects of secondary variables which showed significant effects on one or both of the primary variables were first removed. For assessing the significance of quantitative secondary variables considered, the set of quantitative variables was treated as a set of predictor variables for predicting either SBP or DBP. Multiple Regression Analysis was utilized that identified the subset of significant variables from among the set of predictor variables. Analysis of co-variance was performed in order to examine the effects of categorical secondary variables.

Subjects were identified as hypertensive (SBP values ≥160 mm/ Hg.and/or DBP values ≥95 mm/Hg) and borderline cases (SBP values between 141 and 159 mm/Hg.and/or DBP values between 91 and 94 mm/ Hg) following WHO (1962) criteria. In order to have a conservative estimate, earlier criteria for defining hypertension have been retained. It is but obvious

Table No. 2
Frequency Distribution of Categorical Variables by Gender

Variables	Male		Female		Total	
Education[a]	N	%	N	%	N	%
Illiterate	-	-	5	7.58	5	3.62
Primary	7	9.72	8	12.12	15	10.87
Middle	16	22.22	11	16.67	27	19.57
Secondary	24	33.33	29	43.94	53	38.40
Higher	25	34.32	13	19.69	38	27.54
Oral contra. used	-	-	2	3.03	2	3.03
Disease status[b]						
Hypertension/ Chest pain	3	4.17	3	4.55	6	4.35
Well	69	95.83	63	95.45	132	95.65

a Illiterate: unable to read and write; Primary: 1-5 years of schooling; Middle: 6-8 years of schooling; Secondary: 9-12 years of schooling; Higher: College/University education.

b Hypertension/ Chest pain: reportedly suffering from hypertension and/or have experienced chest pain within 15 days of the survey; Well: not reported to have suffered or suffering from vascular, circulatory, renal disorders or diabetes.

that current WHO criteria, i.e., systole: ≥140 mmHg; diastole: ≥90, if considered, will increase the hypertension prevalence in the study population.

RESULTS

Table No. 3 shows the descriptive statistics [i.e., mean, standard deviation and range of values pertaining to age, SBP and DBP, pulse rate, anthropometric variables including indices, viz., Body Mass Index (BMI) and index of relative subcutaneous fat distribution, smoking, tobacco chewing, alcohol intake and salted tea intake] by gender. The table shows that with respect to both SBP and DBP, males had higher values compared to females, while reverse is true in case of pulse rate. The gender differences, however, are not found to be significant at the 5% level. With respect to anthropometric variables males have higher mean values compared to females except for triceps and sub-scapular skinfold thicknesses as well as for BMI. Further, in respect of numbers per day of cigarettes smoked, tobacco chewed and alcoholic drinks consumed, females generally show lower mean values compared to males; this trend is marginally reversed in case of cups of salted tea consumed, however.

Table No. 3

Descriptive Statistics Pertaining to Quantitative Variables by Gender

Variables	Male				Female				Total			
	N	X	sd	Range	N	X	sd	Range	N	X	sd	Range
Age (years)	72	34.43	12.55	16-69	66	33.45	12.78	16-83	138	33.96	12.63	16-83
Systolic Blood Pressure (mmHg)	72	132.25	16.96	107-188	66	130.02	18.75	91-183	138	131.23	17.81	91-188
Diastolic Blood Pressure (mmHg)	72	91.86	10.72	71-129	66	88.47	11.94	61-119	138	90.24	11.41	61-129
Pulse rate	72	76.47	8.63	60-98	66	81.18	8.70	60-102	138	78.72	8.95	60-102
Weight (kg)	72	57.91	8.69	45.0-80.0	66	55.36	9.61	40.0-90.0	138	56.69	9.20	40.0-90.0
Stature (cm)	72	158.82	5.20	148.3-171.9	66	149.86	5.86	131.0-167.8	138	154.53	7.10	131.0-171.9
Upper arm girth (cm)	72	25.73	2.52	19.0-30.6	66	25.38	3.01	20.6-36.8	138	25.56	2.76	19.0-36.8
Calf girth (cm)	72	32.99	2.31	28.2-37.7	66	32.41	2.54	28.8-41.0	138	32.71	2.43	28.5-41.0
Triceps skinfold (mm)	72	9.13	3.70	4.3-19.8	66	17.55	7.22	5.4-38.0	138	13.16	7.05	4.3-38.0
Subscapular skinfold (mm)	72	16.2	9.44	6.2-42.0	66	27.90	9.84	7.4-48.0	138	21.96	11.16	6.2-48.0
BMI (Kg/m^2)	72	22.96	3.29	17.8-30.5	66	24.64	4.00	18.5-38.3	138	23.76	3.73	17.8-38.3
Index of relative subcutaneous fat distribution[a]	72	0.62	0.74	0.45-0.78	66	0.61	0.07	0.31-0.75	138	0.62	0.07	0.31-0.78
Smoking [b]	6	5.50	3.83	2 - 12	3	3.33	0.57	3 - 4	9	4.78	3.23	2 - 12
Tobacco chewing [c]	21	10.81	5.24	2 - 20	4	7.00	3.83	2 - 10	25	10.20	5.17	2 - 20
Alcohol intake [d]	17	2.06	0.83	1 - 4	11	1.82	0.87	1 - 3	28	1.96	0.84	1 - 4
Salted tea intake [e]	72	2.22	2.10	1 - 10	66	2.55	1.87	1 - 9	138	2.38	1.99	1 - 10

a. Ratio of subscapular skinfold to the sum of triceps and subscapular skinfolds; b. No. of cigarettes smoked per day; c. No. of times tobacco chewed per day; d. No. of alcoholic drinks per day; e. No. of cup of tea, added with salt consumed per day.

The results of the Multiple Regression Analysis using age, anthropometric and behavioural variables as predictors of log transformed SBP and DBP are presented in Table No. 4. Regression analysis is not done separately for the genders since no significant gender effect was noted. Table No. 4 shows that significant quantitative predictors for SBP are biceps girth and the number of tea cups consumed with salt, while tobacco chewing is found to be the only significant quantitative predictor for DBP.

Analyses of co-variance was performed to examine the effects of qualitative categorical variables, i.e., gender, education, oral contraceptive use and disease status on SBP and DBP. Respective significant quantitative predictor variables for SBP and DBP (as shown in Table No. 4) were used as co-variates in each case. The analysis revealed that disease status significantly influences both SBP and DBP (SBP: F ratio = 15.02, d.f. = 1, 134, p = 0.0002; DBP: F ratio = 4.68, d.f. = 1,135, p = 0.0329). Adjusted mean SBP and DBP values remain markedly higher, as expected, among those who suffer reportedly from hypertension and/or chest pain than those who do not suffer from any cardio-vascular, circulatory, renal disorders and diabetes (SBP: 153. 71 mmHg for the diseased group as against 129.11 for the non-diseased group; DBP: 99.36 mmHg for the diseased group as against 89. 11 for the non-diseased group).

The frequency distribution of hypertensive subjects, borderline cases and normotensive subjects by gender is presented in Table No. 5. The table shows that the percentage of hypertensive subjects is conspicuously high in the study population, irrespective of gender. Males show higher percentage of hypertensives compared to females. Again, much higher percentage of hypertensives and borderline cases are thus classified owing to contribution of elevated diastolic pressure only, for both males and females (Table No. 6).

Table No. 4
Results of Multiple Regression Analysis using Age, Anthropometric and Behavioural Variables as Predictors

Dependent Variables	**Sample Size**	**Significant Quantitative Predictors**	**R^2**
Ln SBP	138	Biceps girth, number of tea cups consumed with salt	0.41
Ln DBP	138	Tobacco use	0.37

Both R^2 values are significant at 5% level.

Table No. 5
Frequency Distribution of Categories of Hypertensive Subjects and Borderline Cases and Normotensive Subjects by Gender

Categories	Male		Female		Total	
	N	%	N	%	N	%
Hypertensives	23	31.94	18	27.27	41	29.71
Borderline cases	15	20.83	11	16.67	36	18.84
Normotensives	34	47.22	37	56.06	71	51.45
Total	72	99.99	66	100.00	138	100.00

Table No. 6
Frequency Distribution of Categories of Hypertensive Subjects and Borderline Cases by Gender

Categories	Male		Female		Total	
	N	%	N	%	N	%
Both systolic & diastolic pressures [SBP ≥160 & DBP ≥95]	4	17.39	4	22.22	8	19.51
Systolic pressure only [SBP ≥160]	1	4.35	-	-	1	2.44
Diastolic pressures only [DBP ≥95]	18	78.26	14	77.78	32	78.05
All combined	23	100.00	18	100.00	41	100.00
Borderline cases						
Both systolic & diastolic pressures	3	20.00	4	36.36	7	26.92
Systolic pressure only	1	6.67	1	9.09	2	7.69
Diastolic pressures only	11	79.33	6	54.55	17	65.39
All combined	15	100.00	11	100.00	26	100.00

DISCUSSION

The present micro-level investigation among the Lepcha adults inhabiting Gangtok town and its neighbourhood in Sikkim reveals that the Lepcha do show elevated blood pressure levels. And as a consequence, the prevalence of essential hypertension, as adjudged by the earlier WHO (1962) criteria, is high. It is to be noted that hypertension prevalence becomes still higher in the study group, if World Health Organization-International Society of Hypertension, 1999 criteria for defining essential hypertension (SBP: ≥ 140 mmHg; DBP:≥ 90 mmHg) is considered. It is observed that the prevalence rates for high blood pressure have largely decreased in the developed countries while rates in developing countries continue to rise (Pearson, 1966). Gupta *et al.*'s meta-analysis of data from Indian populations shows an increasing trend in the prevalence of hypertension in urban as well as rural areas. It is also observed that diastolic hypertension was the most common type, as found elsewhere, as well (Harburg *et al.*, 1982). It is surprising to note, however, that unlike many urban or modernizing populations, the present sample of urban – the Lepcha – does not show significant age effects. Many studies, however, observed little or no increase in blood pressure with age in several populations, especially in non-industrialized ones (Cruishank and Beevers, 1985). Significant sex effect is also absent in the study gruop, though males show higher values compared to females for blood pressure.

Of the anthropometric variables considered, biceps girth is found to be a significant predictor of SBP. None of the anthropometric traits is seen to influence DBP, on the other hand. This finding is indicative of the fact that adiposity and/or obesity observed to play important role in influencing blood pressures in many populations, do not influence blood pressure in the population studied. It is revealed that degree of salt intake (as indirectly measured through salted tea intake) significantly predicts SBP in the study population. Cultural preference of adding salt in tea in addition to use of salt in cooked and raw food items, among many of the Lepcha adults, ultimately leads to higher intake of salt which in turn may affect blood pressure level. This finding is in consonance with other studies on the effect of salt intake on blood pressures (Intersalt Cooperative Research Group, 1988). Tobacco chewing is found to be a significant predictor of DBP in the study population. That tobacco use elevates blood pressure is also known (Folsom *et al.*, 1993). It is observed that although smoking is not a major problem in the study population, prevalence of tobacco chewing is reported to be increasing especially among young adults, which may lead to oral as well as cardio-vascular health problems.

As expected, disease status shows significant effect on both SBP and DBP. Previously diagnosed hypertensives or those suffering from chest pain were having significantly higher SBP and DBP. Possibly it may imply that proper medical care is not being taken for the existing blood pressure problems.

Surprisingly, in the study population some important behavioural traits like smoking, alcohol intake and oral contraceptive use do not show any significant effect on blood pressures. It remains difficult to explain such a finding. Alcohol intake, though, significantly influenced blood pressures among the rural Lepchas (Mukhopadhyay *et al.*, 1996).

Although some studies did find association between educational level and blood pressure (e.g., Hutchinson, 1986 and Mukherjee *et al.*, 1986), the present study failed to find any significant effect of education on blood pressure. Dyer *et al.* (1976) had also similar observations.

Multiple factors could be responsible for a generally high level of blood pressure and high prevalence of essential hypertension in the study population. Identification and assessment of relative importance of all such factors are not possible in the present analysis. However, contribution of salt intake and tobacco chewing in influencing SBP and DBP respectively in the study population is evident.

It appears that the urban Lepchas are experiencing some behavioural shifts in adopting dietary changes, leading to higher salt intake and also getting increasingly dependent on tobacco chewing in the changed socio-cultural milieu. Such changes are possibly playing important roles in influencing blood pressure to elevate and hypertensive prevalence to be high. It will be of interest to investigate epidemiologically into the other cardio-vascular risk factors in addition to blood pressure in order to assess cardio-vascular health status among the study population.

ACKNOWLEDGEMENTS

The authors are immensely indebted to the Lepchas of the study area for their help and cooperation; to Prof. P.P. Majumder for his unstinted help in statistical analyses of the data and for valuable suggestions, and to the authorities of the Indian Statistical Institute for financial and administrative facilities.

BIBLIOGRAPHY

Crews, D.E. and S.R.Williams. 1999. 'Molecular aspects of blood pressure regulation'. *Human Biology* 71: 475-503.

Cruishank, J.K. and D.G. Beevers. 1985. 'Ethnic and geographic differences in blood pressure', in C.J. Bulpitt (ed.), *Handbook of Hypertension: Epidemiology of Hypertension*. New York: Elsevier Science Publishers.

Das, A.K. 1978. *The Lepchas of West Bengal*. Calcutta: Editions Indian.

Dressler, W.W. 1999. 'Modernization, stress and blood pressure: New directions in research'. *Human Biology* 71: 583-605.

Dyer, A.J., J. Stamler, R.B. Shekelle and J. Schoenberger. 1976. 'The relationship of education to blood pressure: findings on 40,000 employed Chicagoans'. *Circulation* 54: 987-992.

Epstein, F.H. and R.D. Eckhoff. 1967. 'The epidemiology of high blood pressure: Geographic distribution and etiological factors', in J. Stamler, R. Stamler and T.M. Pullman (eds.), *The Epidemiology of Hypertension*. New York: Grune and Stratton.

Folsom, A.R., K.M. Johnson, H.A. Lando, P.G. McGovern, L.I. Solberg and J.K. Ekstrum. 1993. 'Plasma fibrinogen and other cardiovascular risk factors in urban American Indian smokers'. *Ethnicity and Disease* 3: 344-350.

Gerber, L.M. and R.A. Halberstein. 1999. 'Blood pressure: Genetic and environmental influences'. *Human Biology* 71: 467-473.

Gorer, G. 1938. *Himalayan Village: An Account of the Lepchas of Sikkim*. England: Michael Joseph, London.

Grierson, G. 1927. *A Liguistic Survey of India, v.3, Tibeto-Burman Family*. Calcutta, India: Superintendent of Government Printing.

Gupta, R., N.A. Al-Odat and V.P. Gupta. 1996. 'Hypertension epidemiology in India: Meta analysis of 50-year prevalence rates and blood pressure trends'. *Journal of Human Hypertension* 10: 465-472.

Halberstein, R.A. 1999. 'Blood pressure in the Caribbean'. *Human Biology* 71: 659-684.

Hanna, J.M. and P.T. Baker. 1979. 'Biocultural correlates to blood pressure of Samoan migrants to Hawaii'. *Human Biology* 51: 481-497.

Harburg, E., L. Gleibermann and J. Harburg. 1982. 'Blood pressure and skin color: Maupiti, French Polynesia'. *Human Biology* 54: 283-298.

Hutchinson, J. 1986. 'Association between stress and blood pressure variation in a Caribbean population'. *American Journal of Physical Anthropology* 71: 69-79.

Intersalt Cooperative Research Group. 1988. 'Intersalt: An international study on electrolyte excretion and blood pressure; results for 24-hour urinary sodium and potassium excretion'. *British Medical Journal* 297: 319-328.

Katz, S.H., M. Hediger, J. Schall, E.J. Bowers, W.F. Berker, S. Aurand, P.B. Eveleth, A.B. Gruskin and J.S. Parks. 1980. 'Blood pressure, growth and maturation from childhood through adolescence'. *Hypertension* 2 (suppl. 1): 55-69.

Lohman, T.G., A.F. Roche and R.M. Martorell. 1988. *Anthropometric Standardization Reference Manual*. Champaign, Illinois: Human Kinetic Books.

Lowenstein, F.W. 1961. 'Blood pressure in relation to age and sex in the tropics and subtropics: A review of the literature and an investigation in two tribes of Brazil Indians'. *Lancet* 1: 389-392.

Majumder, P.P., S.K. Bhattacharya, B.N. Mukherjee and D.C. Rao. 1990. 'Genetic epidemiological study of blood pressure in a sedentary rural agricultural population of West Bengal, India'. *American Journal of Physical Anthropology* 81: 563-572.

Majumder, P.P., S. Nayak, S.K. Bhattacharya, K.K. Ghosh, S. Pal and B.N. Mukherjee 1994. 'An epidemiological study of blood pressure and lipid levels among Marwaris of Calcutta, India'. *American Journal of Human Biology* 6: 183-194.

Marmot, M.G. 1979. 'Epidemiological basis for the prevention of coronary heart disease'. *Bulletin WHO*, 57: 331-347.

McGarvey, S.T. and P.T. Baker. 1979. 'The effects of modernization and migration on Samoan blood pressure'. *Human Biology* 51: 461-475.

McGarvey, S.T. and D.E. Schendel. 1986. 'Blood pressure of Samoans', in P.T. Baker, J.M. Hanna and T.S. Baker (eds.), *The Changing Samoans: Behavior and Health in Transition*. New York: Oxford University Press.

Mukherjee, B.N., P.J. Byard, S.K. Bhattacharya and D.C. Rao. 1988. 'Blood pressure in a rural West Bengal fishing community: An epidemiologic profile'. *Human Biology* 60: 69-79.

Mukhopadhyay, B., S. Mukhopadhyay and P.P. Majumder. 1996. 'Blood pressure profile of Lepchas of the Sikkim Himalayas: Epidemiological study'. *Human Biology* 68: 131-145.

Neilson, G., and G. Williams. 1978. 'Blood pressure and valvular and congenital heart disease in Torres Strait Islanders'. *Medical Journal of Australia* (Suppl. 1): 12-16.

Pearson, T.A. 1996. 'Cardiovascular disease as a growing health problem in developing countries: The role of nutrition in the epidemiological transition'. *Public Health Review* 24: 131-146.

Rao, R.S. 1980. 'Blood pressure in relation to age and sex in a semi-urban population'. *Annals of Human Biology* 7: 277-280.

Tamsang, K.P. 1983. *The Unknown and Untold Reality about the Lepchas*. Hong Kong: Luen Sun Offset Printing Co.

Ward, R.H. 1983. 'Genetic and sociocultural components of high blood pressure'. *American Journal of Physical Anthropology* 62: 91-105.

World Health Organization. 1962. 'Arterial Hypertension and Ischemic Heart Disease: Preventive Aspects'. WHO Technical Report Series 231. Geneva: World Health Organization.

21

TRADITIONAL MEDICINE AND HEALTHCARE

S. Gregory

The ability to adapt to any ecological setting has been the hallmark of human evolution. Human beings have been able to spread their wings almost to the entire landmass of the earth and establish their supremacy, overcoming all adverse conditions in their way, long before the development of modern science and technology, solely because of their ingenious ways of adaptation and coping mechanisms that are closely in consonance with the laws of nature. They have evolved appropriate and unique cultural institutions based on their own world of experience, and left a heritage of knowledge for their future generations. Health systems form part of such a cultural milieu of every human society.

Indigenous societies had survived for centuries, overcoming all hazards of physical illness and mental depressions and attained a spiritual ecstasy, through their time-tested remedies, using the bio-resources within their ecosystem, in combination with their propitiation of supernatural elements. The approach had been a holistic one, as their life had been closely integrated with the bio-cultural environment, manifesting a symbiotic relationship and living in consonance with nature. However, with the change of time and their increasing external contacts, they have been experiencing world over, a deteriorating health standard and an ever-increasing threat of physical extinction.

Today, with the intervention and invasion of modern medical systems, a growing dependency has developed among them while they have also been on the verge of losing their rich heritage of knowledge that had been passed on to them from generation to generation. With the world being preoccupied with the claim to Intellectual Property Rights (IPR), and an increasing realization for the use of the bio-resources and indigenous knowledge, the issues of biodiversity, bio-conservation, bioethics, benefit sharing and sustainability had been increasingly raised and addressed in the international fora and conventions.

Developed nations have become developed not under a patent regime. By exploiting the colonies, these countries have become developed. IPR itself is only a recent invention. It could be that to facilitate further exploitation, patenting becomes a must. In this context, it is very important to examine what role could anthropology play in such a regime, and seriously think whether it could provide a value framework in the guidance of humanity.

The present paper takes a look into the implications of the vital concerns raised above regarding the health care system, knowledge and culture of the indigenous people from a humanistic perspective and the impact of the modern forms of invasions on their existence and livelihood.

IPR, FOR WHOSE BENEFIT?

Intellectual property is defined as the creation of human mind, and Intellectual Property Rights (IPR) are legal rights governing the use of such creations (Ekbal, 2006). IPR, which has become part of the International Economic Order, has been conceived to constitute the rights to thoughts, ideas and information regarding new inventions and processes. It enables an inventor to exclude imitators from the market for a specified time. These are rights to make, use and sell a new product or technology that are granted, usually for a period of 17-20 years, solely to the inventor or the corporation which files a claim on the inventor's behalf (Dawkins *et al.*, Net).

How does it get reflected in reality? What kind of impact does it have on the developing nations? Does it lead to the privatisation of common heritage? In spite of legislating biodiversity bills by many countries including India (GOI, 2000), in order to provide for conservation of biological diversity, sustainable use of its components and equitable sharing of the benefits arising out of the use of biological resources and for matters connected therewith or incidental thereto, the concerns still have their place in the present day context.

Exploitative Patent Regime

As per the 1998 estimates, out of four million patents claimed till then world over, only 5 per cent (200,000) of them had come from the developing countries. Even here, the nationals could generate only 30000, a mere 15 per cent, the rest owned by foreign nationals and mainly by the MNCs. Moreover, not even 5 per cent of the foreign-held patents are even used in the productive system of the developing countries. Furthermore, 75 per cent of the world population command just 20 per cent of the world income and 30 per cent of the world trade and hold less than one per cent of the world patent rights. As per the 2004 estimates, 58 per cent of the total patent applications received/granted so far have been owned by the US alone, while with almost one-sixth of the world population, India owns only about 0.1 per cent of the same, clearly indicating whose interest is protected in the patent regime.

Such a situation leads to a series of questions which require serious reflection and response:

- Does the IPR aim at creating global solidarity or producing profitability? If profitability is the aim, which would undoubtedly be so, what is the extent of profit one could justiceably claim?
- What should be the optimal sustainable way of benefit sharing?
- Does it lead to cultural alienation or bio-conservation?
- Does it create an environment conducive to the preservation and promotion of biodiversity or bring about the homogenisation of crop varieties, and mono-cropping, all on monetary terms, and thus destroying the rich biodiversity of the planet earth?
- Does IPR encourage bio-pirating in the traditional heritage, as evidenced by the neem and turmeric episodes, or uphold the validity of the traditional knowledge and customary rights?
- Does it value bio-ethics, celebrate the ownership and encourage capacity-building or promote alienation and disempowerment?
- In short, is not patenting, a latest invention of control mechanism and recolonisation, and IPR, a new ploy of exploitation?
- And finally, is not a strong monopolistic patent regime, unmindful of national interests and scientific roots, an anathema to the humanity itself?

This will do no good except subordinating the universal nature of the laws of commodity production, co-operative and competitive nature of knowledge generation and ownership of common property resources to the monopolistic and restrictive nature of the laws of the IPR regime. In fact, to turn scientific knowledge into a commodity on the market, arguably inhibits science as there are innumerable examples to show that IPRs have been used to suppress innovation (Ekbal, 2006).

Movement of Genetic Resources: Is Compensation Possible?

The movement of genetic resources is not something new as it has always happened in the long history of humanity. Coffee provides an interesting example of the scale of movement of genetic resources during colonial times. Coffee is considered to be native to Ethiopia, transported to Yemen 1000 years ago, and seven seeds were later sent to India, and then to Sri Lanka. The Dutch seem to have taken one plant to Indonesia, and a single cutting to Amsterdom Botonical Garden in 1706. Nine years later, it had been carried to Surinam, and then to Latin America across the Atlantic. Similar is the story with cinchona which revolutionized the Malarial cure, and rubber, which is indispensable of the modern civilization, both discovered in South America. The story could go on for many such bio-resources and their diffusion. How could one value the economic contribution of these innovators or compensate them for their innovation?

The effort to block the outflow of tropical genetic material seemed to have been undermined by the North during and after colonial times. Today the germ-plasm is collected by the MNCs and scientific bodies in the name of bio-prospecting, while very few benefits out of the use of germ-plasm and related knowledge are flown back to local communities. Open access is denied by imposing IPR. In the guise of biological diversity as common heritage of humanity, it is misused by the powerful, who monopolise the resulting financial and technological benefits, through IPR. What is not realized is that the genetic resources of the South are not merely found in nature but survived in nature due to careful nurturing for centuries by local communities and through indigenous innovations. So, if this is common heritage, the same should hold good of the bio-technologies of the North also.

The collection of germ-plasm continues even today, in the name of prospecting by MNCs and scientific bodies. Current economic regimes indicate that the developing countries have massive debts to pay off to the industrialized nations. What these regimes ignore is the far greater debt that the latter had incurred over the last few centuries, in terms of the natural resources, which had flown to them from tropical countries, hardly with any remuneration.

Patenting What? Question of Bio-ethics

Even human genetic material has been up for patenting. The US Department of Commerce had sought a patent on a human cell of a Guayami Indian Woman of Panama of South America, developed from her blood collected by scientists in 1990, resistant to a particular disease and potentially useful in research for AIDS and cancer, leading to a lot of controversies on ethical dimensions (Dawkins *et al.*, Net; Whelan, 2006: 7). However, it had been deferred on ethical protest. Similarly, under the Genome project, there seemed to have been an anthropological hunt for human tissue, hair and blood samples from indigenous peoples of the world, those who are on the verge of extinction.

IPR AND HEALTH CARE

It is perhaps in the field of health that most questions are asked on intellectual property and its role as a promoter of development, as health is a factor that is crucial to the survival and welfare of mankind.

Health Crisis of Humanity

The World Health Organisation (WHO) conceives health as physical, mental and social well being. Health is a product of interaction among socio-economic, political, religio-cultural, biological and environmental factors. It is an essential part of human development, closely related to other fundamental rights such as the right to food/shelter/peace and civic security. Today, the right to life, for the deprived and women, is considered to be more a struggle to buy medicine than to beat the disease.

It has been estimated that the number of very poor people in the world has increased by 10.4 per cent between 1987 and 2001 covering almost half the world population which itself is considered to be an underestimate of the true nature of global poverty (GHW, 2006; Reddy and Pogge, 2006; Satyanarayana and Srivastav, 2007: 391). India, which has emerged as home for holding the largest number of millionaires among the developing countries, also keeps about 77 per cent of its population with an earning of less than twenty rupees (or half a dollar) a day (GOI, 2007). With global health having inextricably linked to the challenges of poverty, the global disease burden is growing steadily with both communicable and non-communicable diseases (Satyanarayana and Srivastav, *ibid.*).

A look at the global health scenario reflects the following facts and figures:

- 1/3 of the global population lack access to potable water and more than a billion suffer from anaemia.
- 2 billion suffer from malnutrition.
- Nearly two million children die every year from diseases for which there is availability of vaccine; 7.5 million children die during prenatal stage, due to lack of care.
- It is reported that diseases like HIV and AIDS, malaria and tuberculosis account for over 90 per cent of the global disease load.
- The female population of developing countries (84% of world's female population) have 1 out of 48 chance of dying during child-birth (in

rich countries, 1 out of 4,000 pregnancies); the indigenous among them constitute a substantial proportion.

- As per the WHO estimates, nearly a third of the world's population lack access to the most basic essential medicines, while in the poorest parts of Africa and Asia, this figure climbs to a half.

In this global scenario of health crisis, India has a massive share. It is in this context that the role of IPR in the health system needs to be viewed.

Unlimited Control of the MNCs

One of the most threatening features of the patent regime is the interference of the MNCs (Multi-National Corporations) in the sovereign rights of the developing nations. It is a trap heavily weighed against the interests of the less developed nations. Before the entry into the patent regime, the plea was that unless we become part of it, we would be left out and isolated. Now that we are already in, and walked into a trap, we cannot escape from it. Many a times, the agreements go against India's basic interests and give the MNCs the full freedom to use legal monopolies to harm the indigenous industries.

Ayurvedic Database and Community/Biodiversity Registers: A Trap or an Immunity?

The CSIR is preparing a database of Ayurvedic drugs to make available to the examiners of all patent offices to attain immunity from further 'neem episodes'. Similar is the case with the biodiversity and community registers (Anuradha *et al.*, 2001). However, there has been a fear at least among some, whether providing such a valuable database to the western world would not open up a vast area of virgin soil for research to the westerners as the MNCs have a competitive edge. While the defenders of the current system might view that it would not be possible to get protection for innovations for knowledge that is already in the public domain, as demonstrated in the 'Basmati' case, the reality is that it involves a high vigilance and the communities hardly have the money to challenge every time a bio-piracy occurs as litigation and similar actions always favour the rich at the expense of the poor (Wild, 2002).

Global Movement against Health Monopoly

Radical amendment has been made of the Indian Patent Act, to conform to the agreement on the Intellectual Property Rights. As a result, there is compulsory protection from 2003, of areas of technology, drastic reduction of price support and food aid to farmers and weaker sections, crippling of scientific and technological potential of developing countries, and weakening of export capacities of the south. The AIDS drugs are highly priced due to patent protection. The cost of treating patients in Africa is said to be many times than that of their GNP, while the same drugs could be produced at one-fourth of its MNC price. On the other hand, while African countries urgently need to increase spending on health care, education and sanitation, not only the IMF structural adjustment programmes have forced these countries to reduce such spending, but have left the debt burden of these countries to increase, taking a lion's share of their GDP (Neiman and Watkins, 1999), leading to more than a double burden on its economy and health scenario. For instance, the sub-Saharan Africa debt rose as a share of GDP from 58% in 1988 to 70% in 1996 (Neiman and Watkins, *ibid.*), forcing the poor countries to divert resources from expenditures on health care and education in order to service external debt.

This has become a rallying point for activists from all parts of the world, resulting in global coalition, and thus creating a momentum of global movement. Brazil/Thailand have defied the TRIPS agreement, and allowed the domestic companies to produce cheap anti-AIDS drug. Though the MNCs and the 'US' had challenged them initially, the WTO had forced to declare allowing governments take independent steps to protect public health, not limited to AIDS alone. However, the steady rise in the disease burden in poor countries, poor access to the existing medicines to those who need them and the near empty pipeline for new drugs for diseases of the poor continue to be one of the major causes of concern in the context of IPR regime (Satyanarayana and Srivastav, 2007: 400).

INDIGENOUS HEALTH SYSTEM

There is not a single indigenous medical system, but as many as there are indigenous cultures. The indigenous medical beliefs and practices are the integral part of a culture and holistic in nature, symbiotically integrated with nature. Under the indigenous system of medicine, ill health is conceived as imbalance in the total system. There are varied types of specialists such as herbalists, diviners, *shaman*, etc.

Invasion of Indigenous Culture and its Impact

There had been physical decimation of the indigenous cultures, by extermination, disease and enslavement. Most of the natives had been killed not by guns but by epidemics. Small pox had been the severest of all the killing diseases, the other ones being whooping cough, measles, leprosy, influenza and veneral diseases. Devastating impact has been reported in Oceania, North America and South Africa.

INDIGENOUS KNOWLEDGE AND IPR – THE CASE OF KANIS

The Kani has become a typical case of discussion and analysis on many aspects in the context of the IPR regime (e.g., Agrawal, Net; Anuradha, Net; Equatoral Initiative, Net; Gupta, 2002; Mashelkar, 2001). The Kani tribe, which inhabits the southernmost Western Ghats region of South India, and which is known to possess a rich health tradition, with potential promises for the community and for the future, is traditionally known to be a nomadic community, who are now settled in the Agasthya forests of Thiruvananthapuram and Kollam districts in Kerala and in the Kanyakumari and Thirunelveli districts of Tamil Nadu. Their population in Kerala is considered to be around 18,000. They live as a group, constituting about 10 to 15 families, in a cluster in the undulated forest terrains. They have limited cultivation around their huts and live a precarious life, mainly depending on the minor forest produce. The development programmes meant for the tribes and the constitutional provisions of independent India have hardly made much of a difference in the lives of the Kanis. The traditional chief of the Kanis is called *Muttukani,* who is said to have lost his authority to a large extent. The local medicine-man, known as *Plathi,* is considered to be the repository of the traditional medical system in the community.

It was in December 1987 that a team of scientists from the Tropical Botanical Garden and Research Institute (TBGRI), led by Dr. Pushpangadan, during an ethno-botanical expedition to the Western Ghats, came across the wonder plant called *arogyapachai.* The two Kani men who accompanied the team as guides, chewing some fruits of a plant, never felt tired while the scientists were feeling fatigued and exhausted during the expedition. The Kani men offered the fruit to the scientists who also felt rejuvenated and energised after consuming the fruit. However, the tribesmen were hesitating to reveal the identity of the plant known as *aragyapachai,* which was a time-honoured secret and a sacred one, though they had to yield at the end, after much persuasion. Further analysis of the plant, involving chemical and pharmacological tests, led to the discovery of the profound adaptogenic and immuno-enhancing properties of the plant, which has been identified as *trichopus zeylanicus travancoricus.*

Based on the research findings, the TBGRI developed a standardized herbal drug, named *jeevani,* including it under the health-promoting group of drugs. Following this, the TBGRI filed a process-patent application in

1996, to obtain an intellectual property protection, which could alone generate greater revenue. *Jeevani* was able to capture the market in India as well as abroad, including countries like USA and Japan (NBRI, 2002). Though the Kani informants were using the fruits of the plant, the preparation of the scientists known as *jeevani* was made using the leaves of the plants. The drug was released for commercial production in 1995 by Arya Vaidya Pharmacy (Agrawal, Net). Nevertheless, it was decided to have a benefit-sharing formula involving the owners of the original knowledge, acknowledging the identical use of the resource and the conservation of biodiversity by the local people that made the invention possible, with transferring the manufacturing right to the Arya Vaidya Pharmacy for a period of ten years for a license fee of ten lakh rupees, with two per cent royalty on future drug sales.

In a local meeting of the Kanis, in order to handle the benefits accrued on account of the drug, it was decided to form a trust under the name 'Kerala Kani Samudaya Kshema Trust', which was registered in November 1997, with nine members, all of whom were Kanis. As per the benefit-sharing formula of 50:50, between the TBGRI and the Kanis, the model of which became the first of its kind in the patent regime (Meshelkar, 2001), the initial share of the license amount of 5 lakh rupees and of the royalties of 19,000 rupees were deposited in the account of Kani Samudaya Kshema Trust at a Union Bank in the locality. In the first meeting of the Trust after the transfer, held in March 1999, it was decided to grant Rupees 50,000 as special incentive to the three informants in the ratio of 2:2:1.

The objectives of the Trust have been stated as:

- Welfare and development activities for Kanis in Kerala
- Preparation of a biodiversity register to document the knowledge-base of the Kanis
- Evolving and supporting methods to promote sustainable use and conservation of biological resources.

The aim of the Trust is to have all adult Kanis in Kerala as its members, though some are opposed to it, either due to lack of knowledge about the Trust or skeptical about its effectiveness of the mode of benefit sharing. Some of the programmes initiated under the aegis of the Trust include setting up a telephone booth, an insurance scheme for pregnant women and to cover accidental deaths, procuring a jeep for transportation along with training a Kani driver, and constructing a community hall.

Issues and Concerns: There are several issues and concerns that emerge from the Kani episode vis-à-vis the patent regime from the point of view of different value orientations such as IPR, CBD, or the different stakeholders like the TBGRI, Forest Department, Tribal Department, KIRTADS, the drug manufacturer, and above all the IK owners themselves. The Kani experience also highlights the lack of coordination and problem-solving mechanisms among the different stakeholders concerning issues of benefit-sharing (Anuradha *et al.,* 2001: 37). There were also accusations and counter-responses with regard to different aspects of the episode (SOS-arsenic.net. 2007; Unnikrishnan, 2004).

As for IPR is concerned, the questions raised are: Was it really a trap or well-intended? Is it a ploy of exploitation by the powerful over the powerless? And if so, is there a way out? Whom does it actually benefit? Is it possible to enter into and explore the positive dimensions of IPR? What are the safety mechanisms? Is it purely materialistic-oriented? Does it aim at creating an equitable and sustainable human society?

As for the CBD, the declaration of the rights of indigenous peoples insists that the knowledge and practices of the local communities are respected and its consent is taken before using such knowledge widely and the resulting benefits are shared with them on an equitable basis. In this context, the question raised could include: Would the hunt for bio-resources lead to the exploitation of biodiversity? Would the new initiatives through IPR help in any way, in the conservation of biodiversity? Are the indigenous communities who had been instrumental in the conservation of local biodiversity given due recognition and encouragement? Was the consent of the local practitioners taken before subjecting their knowledge for further refinement and modification? Are they given due role of participation in the process of refinement of the knowledge production? To what extent are the knowledge and practices of the local communities respected?

On the part of the TBGRI: Having evolved a model, which has come to be known by different names as the 'Kani Model', 'TBGRI Model' or 'Pushpangadhan Model' of benefit sharing (NBRI, 2002), even before the signing of the CBD, could there be a better way than the one adopted under the given circumstances and environment, in spite of the absence of the legal requirement for TBGRI to share the commercial benefits with the Kanis? Is there any other model that has been developed with greater acceptability than the one developed by them?

On the part of the Forest Department: Why were they not involved from the beginning of the process? Should the tribes be given the right to collect the plants to supply for commercial use? Would the mass cultivation of the plant affect the conservation of the particular species?

On the part of the Kanis themselves: Is there a consensus among all in parting the traditional knowledge to outside society? What is the extent of their claim to the knowledge? How should the benefit be shared? Should the new medicinal product be a must for the kind of development plans being initiated now, utilising the funds raised through benefit-sharing arrangement? How wide is their participation in the decision-making process? Are the traditional institutions given due recognition in the process? Are there any groups or sections sidelined? What role are Kani medicine-men given in the entire process? Has the concept of development for the Kanis changed today? Is it different from the one in the past? There are contradicting views and perceptions in the community with regard to all these aspects, truly reflecting the chaos and confusions surrounding the episode in particular and the functioning of the patent regime itself in general.

The other general concerns raised in different forums include: Is not the contracted license fee too low to match the economic potential of the product? Is not the contribution of the Kanis undervalued? Should not the licensing be given to a public sector company rather than to a private company? Is it an instance of bio-piracy as it is claimed by some?

According to KIRTADS, the institute that is concerned with the research, training and development of the Scheduled Castes and Scheduled Tribes in Kerala, the Kanis have got an unfair deal and they were not involved in the negotiating process. It had sought powers to screen all those who approach Kerala's tribal healers in connection with ethno-botany, ethno-medicine and other research in which tribal knowledge is at stake. It has also expressed concern over the preservation of tribal medicine in its original form. A proper co-ordination mechanism among the different stake-holders, as suggested by some studies, should be the need of the hour in addressing some of the concerns mentioned above.

All critical issues connected with Kani-TBGRI experiment are relevant in most analogous situations involving the rural (and indigenous) communities of Kerala (Unnikrishnan, 2004: 44). In this context, a host of complexities also need to be resolved, concerning the ownership of the resources and related knowledge, especially when such resources occur on private land, public land or both types of land, and when the resources and related knowledge occur in more than one (community or) state (Anuradha *et al.*, 2001: 38).

The Government of Kerala in its Intellectual Property Rights Policy proposes to establish 'knowledge commons', 'commons licence' and 'creative commons' (GOK, 2008; PRAGOTI, 2008) in order to protect the

Intellectual Property Rights of the indigenous communities, with the intention to provide a legal status to such knowledge and to prevent the private misappropriation.

AN AGENDA FOR FUTURE

We are living in a world of contradictions; we are in a confused state. In that sense, everyone of us is a layman, unsure of what is right and what is wrong, though we are guided by our own convictions and ideological leanings. But then, there can be certain criteria as the following, which may help us judge what is desirable and what is not in taking a reasonable risk.

- History could be one such indicator: Do we have any lesson to learn from the history of colonisation?
- The talisman of Gandhiji: Whenever you are in doubt or when the self becomes too much with you, recall the weakest person whom you may have seen and ask yourself: If the step you contemplated is going to be of any use to that person? Will that person gain anything by it? Will it restore to that person a control over one's own life and destiny?
- What Franklin D. Roosevelt said: The test of our progress is not whether we add more to the abundance of those who have much; it is whether we provide enough for those who have little.
- By asking an ethical question: Who owns and who benefits? What kind of sharing? Are the owners empowered or alienated?
- And a value-based question: Which is better – materialistic or humanistic value?

BIBLIOGRAPHY

Agrawal, D.P. Net. 'The *Jeevani* elixir of the Kani tribes of Kerala and their Intellectual Property (IP) rights'. http://www.infinityfoundation.com/mandala/t_es/t_es_agraw_jeevani_frameset.htm.

Anuradha, R.V. Net. 'Sharing with the Kanis: A Case Study from Kerala, India'. http://www.cbd.int/doc/case-studies/abs/cs-abs-kanis.pdf.

Anuradha, R.V., Bansuri Taneja and Ashish Kothari. 2001. 'Experiences with Biodiversity Policy-Making and Community Registers in India. Participation in Access and Benefit-Sharing Policy, Case Study No. 3'. London: IIED. https://www.cbd.int/doc/case-studies/abs/cs-abs-reg-in-en.pdf.

Dawkins, Kristin, Michelle Thom and Carolyn Carr. Net. 'Intellectual Property Rights and Biodiversity'. http://ces.iisc.ernet.in/hpg/cesmg/iprdoc.html.

Ekbal, B. 2006. 'Intellectual Property Rights: Challenges to Academic Research'. http://www.sfikeralaejournal.org/ekbal_20062008.htm.

Equatoral Initiative. 2002. http://www.equatorinitiative.net/files/2002-0184_Nom_KeralaKaniTrust_India.pdf.

GHW. 2006. *Global Health Watch 2005-2006: An alternative world health report*. London: Zed Books.

GOI. 2000. 'Biological Diversity Bill'. http://www.grain.org/brl_files/india-biodiversityact-2002.pdf.

GOI. 2007. *Report on the conditions of work and promotion of livelihoods in the unorganized labour.* New Delhi: National Commission for Enterprises in the Unorganised Sector (NCEUS), Govt. of India.

GOK. 2008. 'Draft Intellectual Property Rights Policy for Kerala 2007'. http://www.pragoti.org/sites/default/files/Kerala%20IP%20policy%20draft_0.pdf.

Gupta, Anil K. 2002. 'Value addition to local Kani tribal knowledge: patenting, licensing and benefit sharing'. *IIMA Working Paper*. No. 2002-08-02, August 2002. http://www.iimahd.ernet.in/publications/data/2002-08-02AnilKGupta.pdf.

——. 'Conserving Biodiversity and Rewarding Associated Knowledge and Innovation Systems: Honey Bee Perspective'. http://www.hks.harvard. edu/sustsci/ists/TWAS_0202/gupta_210999.pdf.

Mashelkar, R.A. 2001. 'Intellectual property rights and the Third World'. *Current Science* 81 (8): 955.

Naiman, Robert and Neil Watkins. 1999. 'A Survey of the Impacts of IMF Structural Adjustment in Africa: Growth, Social Spending, and Debt Relief'. April 1999, Centre for Economic and Policy Research. http://www.cepr.net/index.php/a-survey-of-the-impacts-of-imf-structural-adjustment-in-africa/#1.

NBRI. 2002. 'Pushpangadan P Model of benefit sharing'. http://www.nbri-lko.org/director%20data/index10.htm.

PRAGOTI. 2008. 'Kerala Proposes "Knowledge Commons"'. http://www.pragoti.org/node/1564.

Reddy, S. and T. Pogge. 2006. 'How not to count the poor', in S. Anand, J. Stiglitz (eds.), *Measuring global poverty*. Oxford: Oxford University Press, 2006.

Satyanarayana, K. and S. Srivastava. 2007. 'Poverty, Health and Intellectual Property Rights with Special Reference to India'. *Indian J. Med. Res.* 126, October: 390-406.

SOS-arsenic.net. 2007. 'Biodiversity-India: Tribe Accuses Biologists'. http://www.sos-arsenic.net/english/homegarden/biologists.html.

Unnikrishnan, E. 2004. '*Materia Medica* of the Local Health Traditions of Payyannur'. Discussion Paper No. 80. Kerala Research Programme on Local Level Development. Centre for Development Studies. Thiruvananthapuram. www.krpcds.org/publication/downloads/80.pdf.

Whelan, Marina L. 2006. 'What, if any, are the Ethical Obligations of the U.S. Patent Office? A Closer Look at the Biological Sampling of Indigenous Groups'. *Duke Law and Technology Review* No. 14. http://www.law.duke.edu/journals/dltr/articles/pdf/2006DLTR0014.pdf.

Wild, Joff. 2002. 'The Future for Patents on Life'. http://scientific.thomsonreuters.com/free/ipmatters/patlife/8180006/.

22

COMMUNITY AND TUBERCULOSIS

Salil Kumar Dutta

WHAT IS TUBERCULOSIS?

Tuberculosis is an infectious disease caused by tubercle bacilli *(Mycobacterium tuberculosis)*. The disease primarily affects lungs and causes pulmonary tuberculosis. It can also affect intestine, meninges, bones and joints, lymph glands, skin and other tissues of the body. The disease also affects animals such as cattle; this is known as 'bovine tuberculosis' which may sometimes be communicated to man. Pulmonary tuberculosis, the most important form of tuberculosis which affects man, is presently given more attention.[1]

WHY MORE ATTENTION?

India accounts for about 33% global burden of tuberculosis. The disease is one of India's most important public health problems. Everyday in India more than 20,000 people become infected with tubercle bacillus, more than 5000 develop the disease, and over 1000 die from TB. In India, tuberculosis kills 14 times more people than all tropical diseases combined, 21 times more than malaria, and 400 times more than leprosy. Every year, nearly 20 lakh people develop tuberculosis in India, nearly one million of them are highly infections sputum-positive cases. Two such cases develop every minute.

Every sputum-positive patient can infect 10-15 individuals in a year. Tuberculosis also kills more women than all causes of maternal mortality combined. HIV and multi-drug resistant TB threaten to make this situation even worse. Unless urgent action is taken, more than 40 lakh people in India will die in the next decade.

Chart No. 1 shows how TB patients are growing day by day in each and every quarter of the year. The RNTCP programme is detecting the cases and treating with DOTS, where nearly 50,000 patients are increasing in each quarter, though the programme has yet to reach in every corner of the country.

Chart No. 1
Population in India Covered under DOTS and Total Tuberculosis Patients put on Treatment Each Quarter

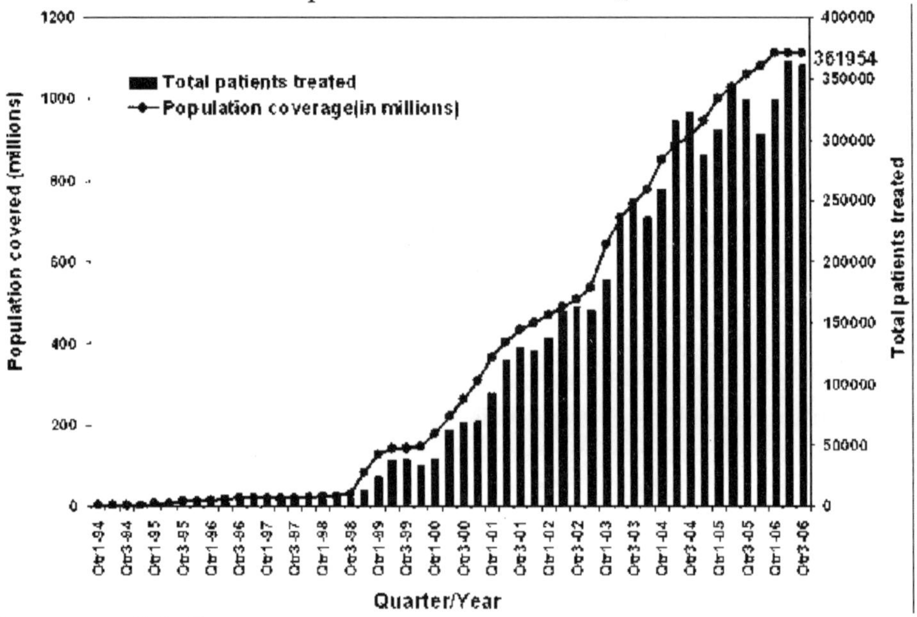

Source: www.tbcindia.org

Combating TB, India has long been at the forefront of global battle against tuberculosis. Pioneering studies at Tuberculosis Research Center, Chennai and the National Tuberculosis Institute, Banglore established the principles of tuberculosis control which are followed even to this day the world over. These principles include the effectiveness of ambulatory treatment of tuberculosis, the effectiveness of intermittent treatment regimens. The necessity of direct observation of treatment by a trained individual who is not a family member, the usefulness and practicability of microscopy as a diagnostic tool among patients reporting to health and the crushing burden of tuberculosis on our society.[2]

Diabetics run high TB risk: Diabetes has now found to be fuelling India's deadly TB epidemic. It may be the primary reason behind more than 10% of tuberculosis in India and China. TB occurs more often among people with diabetes than in those without. So people with diabetes may be important target to regularly screen for treating active TB in areas of high TB incidence.

Epidemiological factors[3]

Agent: Of importance to man is the 'human strain' of *Mycobacterium Tuberculosis*, which is responsible for the vast majority of cases. The 'bovine' strain affects cattle and other animals.

Sources of infections: There are two sources of infection – human and bovine.

(i) *Human source*: The main source of infection is the patient whose sputum is positive for tubercle bacilli.
(ii) *Bovine source*: The bovine source of infection is usually infected milk and milk products.

Age: Tuberculosis can occur at any age. According to RNTCP status Report 2006, it is mainly the working group of population, i.e., 15-54 years age group.

Sex: Males get infected more than females.

Social Factors: Tuberculosis is more likely to occur in persons

(i) who are malnourished,
(ii) who live in overcrowded houses with infected people, and
(iii) where standards of hygiene are poor
(iv) Diabetics run high TB risk.

Period of infectivity: This disease is infectious as long as the bacilli are excreted in the sputum by the human host. This may be from several months to a few years if the case is not adequately treated.

Mode of transmission:

a) *Droplet infection*: TB is spread mainly by droplet infection by an infectious case. Coughing generates the largest number of droplets.
b) *Other ways*: Pulmonary TB is also transmitted by inhalation of infected dust.

Incubation period: This may be weeks or months, depending upon the host parasite relationship and the dose of infection.

Who are TB suspected?

The commonest signs and symptoms which are likely to bring the patient to medical care institutions are:

a) cough which lasts for three weeks or more
b) low grade fever, especially with rise in temperature in the evenings,

c) chest pain,
d) haemoptysis,
e) loss of weight,
f) tiredness,
g) night sweats,
h) shortness of breath and
i) loss of appetite.

Diagnosis of TB

All patients in out-patient departments complaining for cough for 3 weeks or longer are directly sent to designated microscopy centre (DMC) for acid-fast bacilli test (AFB) for three sputum sample of spot-morning and spot collection system. If any two of those smear results are found positive, they are diagnosed as smear positive case. Based on their treatment history, they are further classified as new or previously treated cases (Relapses, Failures, Treatment after default cases) and appropriate treatment regimen is prescribed.

Patients with only one positive smear result are referred to the nearest X-ray facility. Of these, patients who have chest X-ray compatible with TB as diagnosed by a medical officer (MO) are considered to be suffering from TB and registered as smear positive cases.

Patients with three negative results are termed as smear negative and are prescribed broad spectrum antibiotics for 1-2 weeks. If the symptoms persist, the patient is re-evaluated on the basis of X-ray and clinical examinations.[4]

Recommendation of Anti-TB Treatment (RNTCP)

The National Tuberculosis Control Programme launched in 1962, has been modified recently (1992) with efforts to rectify the previous lacunae and incorporating the World Health Organization (WHO) guidelines of treatment. The approach starts with defining a case of tuberculosis with the determination of the site of improvement, severity of disease, bacteriological status and the history and details of previous treatment and subsequent categorization of the patients for treatment guideline.[5] The targets are to detect at least 70% of new smear positive (NSP) cases, offering short-course treatment to all of them under direct observation (through DOTS) and to achieve at least 85% cure rate. It is estimated that approximately 750 NSP cases per million population are to be treated in this programme.[4]

Standardized treatment regimens are recommended by WHO for each category of treatment. These recommended treatment regimens are proven to be effective. The treatment for TB under the DOTS strategy is divided into two phases: intensive and continuation phases. Direct observation of treatment is recommended for all patients. During the intensive phase, each dose has to be directly observed. During the continuation phase, at least the first of the three weekly doses should be given under direct observation.

Treatment Regimens under the RNTCP

Category of Treatment	Type of patient	Regimen
Category I	New sputum smear-positive Seriously ill sputum smear-negative Seriously ill extra pulmonary	2(HRZE)3/ 4(HR)3
Category II	Previously treated Sputum smear-positive Relapse Sputum smear-positive Failure Sputum smear-positive Treatment After Default	2(HRZES)3/ 1(HRZE)3/ 5(HRE)3
Category III	New sputum smear-negative, not seriously ill extra pulmonary, not seriously ill	2(HRZ)3/ 4(HR)3

H – Isoniazid 600 mg. R – Rifampicin 450 mg. Z – Pyrazinamide 1500 mg. E – Ethambutol 1200 mg. S – Injection Streptomycine 750 mg.

Direct observation of treatment has emerged as the standard of care in developed as well as developing countries. Forty years ago, studies conducted at the Tuberculosis Research Center, Chennai provided empirical evidence of the necessity and feasibility of directly observed treatment for achieving a high cure rate for TB. More recently, a study conducted in Pathanamthitta district, Kerala demonstrated that the probability of failure or relapse was 15 times higher among patient who did not receive directly observed treatment as against those who did. It is vitally important for the whole community that people with TB take all their medications on schedule. Interrupted treatment results in chronically infectious cases of TB, some of whom may develop multi-drug-resistant TB.[1]

DIRECTLY OBSERVED TREATMENT SHORT-COURSE (DOTS)

DOTS is a five-point strategy. It includes diagnosis primarily through microscopy, uninterrupted supply of good quality drugs, direct observation of treatment, and monitoring and supervision to track diagnosis, progress and outcome of TB.

Methods

To study the problems of tuberculosis suffering patients and their treatment in an isolated island purposively selected Rudranagar Tuberculosis Unit (Sagar and Namkhana blocks) in South 24 Parganas of West Bengal. To know about their socio-economic condition, the researcher participated with the community more than 5 years; even he actively participated with treatment supervisory in the community for his study. Fifty case-studies, (NSP) genealogical studies and 50 focus-group interviews were conducted to study the problems of the patients and community.

To analyse data simple percentage, graphs and charts with tables were used by the researcher. To collect the name and addresses of TB patients, he consulted the Tuberculosis Register, Laboratory Register and met with medical officer for TB control (MOTC) and other medical officers for referral system of TB suspected patients in Several Peripheral Health Institutions, Block Primary Health Centres (BPHC), DOTS Providers, Health Assistants in Sub-Centre level and community level directly as a Senior Treatment Supervisor (STS). The researcher also consulted with libraries and websites like www.tbcindia.org etc.

Objectives

The present study intends to delineate the problems and prospects of RNTCP/DOTS in the community in a Tuberculosis Unit, which is isolated from the main land in western Sunderban area of south 24 Parganas district of West Bengal.

HOW RNTCP IS ORGANIZING IN COMMUNITY LEVEL

It is a National Programme. Every state is taking initiatives to organize the programme. They have formed a society, namely, State Health and Family Welfare Society, of which each district of the State is responsible for carrying on the programme. The District Health and Family Welfare Society are organizing and implementing the programme. They have few special staffs to implement properly, which are trained by the RNTCP guideline. The team consists of:

1. District Tuberculosis Officer (DTO)
2. Medical Officer for Tuberculosis Control (MOTC)
3. Senior Treatment Supervisor (STS)
4. Senior Laboratory Supervisor (STLS)
5. Laboratory Technician (LT)
6. X-Ray Technician (MT-X-ray)

7. Pharmacist
8. Health Assistants (HA)
9. Community level DOTS provider

MOTC is the head of the Tuberculosis Unit (TU). A Tuberculosis Unit consists of five lakh population with five Designated Microscopic Centre (DMC) and all other staffs given above. DTO and STO with a WHO consultant are monitoring the programme each and every moment.

ABOUT SUNDERBAN AND RNTCP

Sunderban, an active delta of India, consists of 19 blocks of which 16 are in South 24 Parganas and 3 in North 24 Parganas districts in West Bengal. All the area started with RNTCP/DOTS programmes from August 2000. But the part of South 24 Parganas district has been started from March 2001. The Western Sunderban eco-region is our study area which is situated on the eastern bank of the river Hooghly (particularly Sagar and Namkhana, these two blocks are under Rudranagar TU).

Table No. 1 shows the annualized case detection rate in district South 24 Parganas, West Bengal, where the population covers all corners of the district (73000000 approx.), but the rate is more or less stationary. Cure and conversion rates are always the target, whereas the death rate is always below the target which seems the perfection of the programme.

ABOUT RUDRANAGAR TU

Rudranagar Tuberculosis Unit is one of the 15 units in South 24 Parganas district of West Bengal, situated at the extreme southernmost part of the state. This again consists of two blocks – Sagar and Namkhana – with its 5 islands in Western Sunderban eco-region in the northern bank of the Bay of Bengal and eastern bank of the river Hooghly. The population of this TU is 3,70,250 (2001 census), area approximately 500 sq. km. having 14 Medical Officer (MO), STS, STLS and LTs with 4 DMC and 79 sub-centres. Besides this, more than 20 Rural Medical Practitioners (RMP) (locally known as quack) are involved in this programme for carrying on work in several remote corners of the island.

Prospects of RNTCP in Rudranagar TU

Since March 2001 this programme has been implemented and all items of the infrastructure are working properly. Sufficient medicines and monitoring regularly making upgraded to this programme.

Table No. 1
Performance Report of RNTCP South 24 Parganas for the Year 2003 to 3rd Qtr. 2006

Total & Annualised Case Detection				Total N.S. Pos. & N.S. Pos Detection Rate				Ratio of N.S. Pos & Negative		% of N.S.POS Put on DOTs	
TARGET-Annualised Case Detection Rate 142 per lakh per year(70% of 203 Per lakh/year)				TARGET-Annualised N.S. POS Detection Rate 53 per Lakh per year (70% of 75 per Lakh/year)				TARGET- 1:1.2		TARGET-90% of NSP Detected	
Quarter	Total Case	A.case D.R. & %		Quarter	Total Case	N.S.Pos D.R & %					
1st Qtr, 03	1396	78	38.4%	1st Qtr, 03	582	32.6	43.31%	1st Qtr, 03	0.8	1st Qtr, 03	93%
2nd Qtr., 03	1399	78	38.4%	2nd Qtr., 03	614	34.4	45.86%	2nd Qtr., 03	0.7	2nd Qtr., 03	90%
3rd Qtr., 03	1381	77	37.9%	3rd Qtr., 03	538	30	40%	3rd Qtr., 03	0.98	3rd Qtr., 03	94%
4th Qtr., 03	1290	72	35.46%	4th Qtr., 03	496	27.8	37.08%	4th Qtr., 03	0.8	4th Qtr., 03	92%
1st Qtr., 04	1481	81	40.%	1st Qtr., 04	693	38.2	50.93%	1st Qtr., 04	0.5	1st Qtr., 04	94%
2nd Qtr., 04	1611	88.8	43.7%	2nd Qtr., 04	786	43.3	52.72%	2nd Qtr., 04	0.43	2nd Qtr., 04	94%
3rd Qtr., 04	1693	93	45.8%	3rd Qtr., 04	793	43.7	58.7%	3rd Qtr., 04	0.52	3rd Qtr., 04	94%
4th Qtr., 04	1738	95.8	47.1%	4th Qtr., 04	869	47.9	63.8%	4th Qtr., 04	0.46	4th Qtr., 04	95%
1st Qtr,. 05	1676	90.87	44.7%	1st Qtr,. 05	817	44.29	59.25%	1st Qtr,. 05	0.45	1st Qtr,. 05	98%
2nd Qtr., 05	2009	109	53.6%	2nd Qtr., 05	978	53	70.7%	2nd Qtr., 05	1:0.48	2nd Qtr., 05	93%
3rd Qtr., 05	1856	101	50%	3rd Qtr., 05	958	52	69.3%	3rd Qtr., 05	1:0.42	3rd Qtr., 05	96%
4th Qtr., 05	1861	101	50%	4th Qtr., 05	962	52.16	69.5%	4th Qtr., 05	01:00.4	4th Qtr., 05	95%
1st Qtr., 06	2156	117	57.63%	1st Qtr., 06	1081	59	78%	1st Qtr., 06	1:0.4	1st Qtr., 06	95%
2nd Qtr., 06	2014	109	53.69%	2nd Qtr., 06	1049	57	76%	2nd Qtr., 06	1:0.37	2nd Qtr., 06	96%
3rd Qtr., 06	1847	100	49.26%	3rd Qtr., 06	956	52	69%	3rd Qtr., 06	1:0.36	3rd Qtr., 06	97%

3RD Month Conversion Rate		Cure Rate		Default Rate		Death Rate	
TARGET - 90%		TARGET - 85%		TARGET - 5%		TARGET - 4%	
4th Qtr, 02	91%	1st Qtr, 02	86%	1st Qtr, 02	8.11%(86)	1st Qtr, 02	3.68%(39)
1st Qtr., 03	91%	2nd Qtr., 02	87%	2nd Qtr., 02	5.21% (59)	2nd Qtr., 02	3.71%(42)
2nd Qtr., 03	91%	3rd Qtr., 02	88%	3rd Qtr., 02	6.05%(70)	3rd Qtr., 02	4.41% (51)
3rd Qtr., 03	89%	4th Qtr., 02	87%	4th Qtr., 02	5.68% (84)	4th Qtr., 02	4.26%(54)
4th Qtr., 03	92%	1st Qtr., 03	88%	1st Qtr., 03	5.41% (63)	1st Qtr., 03	3.95%(46)
1st Qtr., 04	91.2% (632)	2nd Qtr., 03	86.8% (533)	2nd Qtr., 03	5% (31)	2nd Qtr., 03	4.31% (27)
2nd Qtr., 04	93% (731)	3rd Qtr., 03	88.84% (478)	3rd Qtr., 03	5.39% (29)	3rd Qtr., 03	2.6% (14)
3rd Qtr., 04	93.3% (740)	4th Qtr., 03	88.9% (441)	4th Qtr., 03	5% (25)	4th Qtr., 03	2% (10)
4th Qtr., 04	90.44% (786)	1st Qtr., 04	87.01% (603)	1st Qtr., 04	6.2% (43)	1st Qtr., 04	2.59% (18)
1st Qtr., 05	89% (757)	2nd Qtr., 04	87% (682	2nd Qtr., 04	5.85% (46)	2nd Qtr., 04	3.05% (24)
2nd Qtr., 05	93%(909)	3rd Qtr., 04	90% (716)	3rd Qtr., 04	4.79% (38)	3rd Qtr., 04	2.39% (19)
3rd Qtr., 05	91%(874)	4th Qtr., 04	88% (769)	4th Qtr., 04	5.86% (51)	4th Qtr., 04	2.99% (26)
4th Qtr., 05	93%(890)	1st Qtr., 05	87% (707)	1st Qtr., 05	6.2% (51)	1st Qtr., 05	4.16% (34)
1st Qtr., 06	92%(991)	2nd Qtr., 05	91% (890)	2nd Qtr., 05	4.19% (41)	2nd Qtr., 05	2.76% (27)
2nd Qtr., 06	92%(969)	3rd Qtr., 05	89% (850)	3rd Qtr., 05	6.15% (59)	3rd Qtr., 05	2.5% (24)

Note: Caculation done as per projected population of 2003 (71.36), 2004 (72.56), 2005 (7377839) & 2006 (7387799).

Source: DTO South 24 Parganas.

Table No. 2 reveals year-wise status of Rudranagar Tuberculosis Unit since 2001, where the case detection is poor but the success rate has reached above the target.

The above table shows, how from 2001 to 2006 case findings are improving year-wise. In the 1st year it was 74 cases only, whereas in 2006 it reached up to 358 cases. So, it is more or less 5 times increase from the case findings of 2001. Still the target is far behind expectation. In every category it is being exhibited that the females are lesser in number than males; mainly in case of new smear-positive cases, it is more than two times higher than the females.

Table No. 2
Performance Report of Rudranagar TU from 2nd Quarter, 2001 to 4th Quarter, 2006

Period	Population (in lakh)	% of out-patients Examined	% S+ve living in T.U. put on DOTS	Total Cases Treated	Annualized Total Detection Rate	New S+ve Cases Treated	Annualized Smear +ve Detection Rate	Ratio of S -ve to S +ve Patients	3 Months Conversion rate of New S +ve	Cure rate of new Smear +ve Patients	Success rate of New S +ve Patients
2001											
2nd Qtr., 01	3.46	0.50%	100%	35	40	11	13	0.9	-	-	-
3rd Qtr., 01	3.46	1.00%	93%	31	36	11	13	0.7	64%	-	-
4th Qtr., 01	3.46	2.00%	100%	26	30	7	8	1.1	82%	-	-
2002											
1st Qtr., 02	3.46	3.10%	100%	33	38	12	14	0.41	100%	-	-
2nd Qtr., 02	3.46	2.10%	91%	43	50	23	34	0.3	92%	55%	55%
3rd Qtr., 02	3.46	2.86%	92%	38	44	19	22	0.5	87%	82%	82%
4th Qtr., 02	3.46	2.12%	96.29%	43	49.71	19	21.97	0.4	89%	100.00%	100.00%
2003											
1st Qtr., 03	3.46	2.09%	96%	35	40	22	25	0.1	90%	92%	92%
2nd Qtr., 03	3.46	1.99%	88%	43	49.71	21	24.27	0.61	100%	78%	78%
3rd Qtr., 03	3.46	3.00%	83%	29	34	11	13	0.7	86%	84%	84%
4th Qtr., 03	3.46	1.89%	100%	33	38	21	24	0.1	94%	91%	91%
2004											
1st Qtr., 04	3.46	1.89%	100%	33	38	21	24	0.1	94%	91%	91%
2nd Qtr., 04	3.46	1.38%	90%	37	42.7	19	21.00	0.3	100%	78%	78%
3rd Qtr., 04	3.46	1.50%	94.40%	53	61	31	35.83	0.3	79%	82%	82%
4th Qtr., 04	3.46	1.93%	100.00%	50	58	29	34	0.41	97%	94%	94%
2005											
1st Qtr., 05	369755	2.05%	100.00%	36	38.94	20	21.63	0.35	100%	76%	76%
2nd Qtr., 05	3.69	1.75%	100%	40	43	25	27	0.28	80%	74%	84%
3rd Qtr., 05	3.69	1.99%	100%	61	66	30	33	0.43	100%	94%	94%
4th Qtr., 05	3.69	2.66%	100%	76	82	40	43	1:0.52	87%	97%	97%
2006											
1st Qtr., 06	3.70	3.53%	100%	83	90	39	42	1:0.7	88%	85%	85%
2nd Qtr., 06	3.70	3.10%	100%	103	111	42	45	1:0.73	97%	100%	100%
3rd Qtr., 06	3.70	2.97%	100%	69	75	38	41	1:0.31	95%	90%	90%
4th Qtr., 06	3.70	3.10%	100%	100	108	52	56	1:0.28	97%	83%	88%

Source: DTO South 24 Parganas.

Table No. 3
Sex-wise Distribution of Patients in Rudranagar Tuberculosis Unit since 2001

Patients Category	NSP		RT		NSN		EP		TOTAL	
Year	Male	Female	Male	Female	Male	Female	Male	Female	Male	Female
2001	16	7	18	7	11	8	3	4	48	26
2002	45	17	22	10	15	22	3	1	85	30
2003	47	12	12	6	12	4	5	4	76	26
2004	76	23	20	4	21	10	10	9	127	46
2005	79	39	12	5	30	15	20	11	141	70
2006	116	55	27	8	57	30	39	26	239	119
TOTAL	379	153	111	40	146	69	80	55	716	317

Source: TB Register Rudranagar TU.

Treatment outcome is also very good; it is always 85% in every year in case of NSP. Patients are taking medicines and completing treatment as more than 80% are found in Tuberculosis Register.

Few cases (5 cases in last 7 years) are failures and they were suspected with HIV positive. They all died within or after their treatment period. All of them have strong history of multi-partner sex habits. They once worked with construction group at several parts in our country.

WHAT ARE THE PROBLEMS?

There are several problems found out from more than 50 focus group interviews from both sexes and several working groups and age groups. The results are given below:

I. The medicine is very powerful and it harms the body; as per 70% of the people.
II. 50% of the people told: injection may be better than this lot of drugs.
III. Prolonged exposure to medicine hamper daily life, told 90% of the people.
IV. During DOTS it creates social problems, like unability to attend daily works and social occasions. Most of the Muslims, above 90%, do want to stop medicines during their fasting days (Ramzan month).
V. The day they swallow medicines, it means hampering of all activities even having sex too, which is intolerable; told almost all males of above 30 years to below 55 years.
VI. As it takes prolonged treatment with DOTS, many people identify even the unmarried members of the family. This creates a massive problem to arrange marriage. Case-study of Ms. R. and Ms. M.K., the two government servants expressed their story of life (the original names are not published due to unavoidable social circumstances).
VII. DOTS hamper seasonal business, which involves migration from one place to another. A patient was cured after 2nd time of Cat-II regiment, though he was started with NSP-Cat-I regiment.
VIII. DOTS disturb the business of private practitioners, so the patients are de-motivated.
IX. Sometimes patients avoid government assistance due to social fear. They are used to go to private practitioners very sincerely and take medicine at their home. Hence, private practitioners take the advantages.
X. As fishing is one of the backbone of the island economy (nearly above 60% of costal people are live on fishing in the sea), so application of DOTS is to some extent impossible; told more than 60% of the people

attending the focus-group discussions. The patients are always delayed to complete the doses.

XI. Maximum DOTS providers are government health assistants, who are so much over-burdened by several vertical programmes of health department that sometimes they cannot do DOTS properly; told 85% of the health assistants in their focus group discussions.

XII. Patients with smear-positive are asked to stay at a distance in mother and children's clinic at sub-centre level. This damages the moral of the patients, said 85% of RT patients and 63% of the NSP patients and 35% of the NSN patients during their in-depth interviews.

FEW CASE STUDIES

Case-1: Unfortunate wise women

I, Ms. R. of Namkhana, age 24 years, unmarried graduate, had come to hospital with the symptoms of cough for one month and low fever with loss of appetite with chest pain on March, 2003. The doctors diagnosed me as a smear-positive patient. Previously, I never took medicine for Tuberculosis. So I was asked to take medicine for six months. STS told me about the medicine which is category-1. I got infected when I was working as a nurse at a nursing home in Kolkata. I was requested by STS to hand over full drugs to my father who is an educated farmer. But STS tried to follow the guideline of RNTCP and motivated me to take medicine from a person who is a school teacher. I regularly took medicines and ultimately got cured within 6 months. But as the matter was known to everybody, nobody wants to marry me. Presently I am working as a school teacher and still I am unmarried. Tuberculosis cured my lungs but broke my heart.

Case-2: Punishment of truth and chastity

I, Ms. M.K. am working in a government health system since 1990. I am 38 years in age. I was treated with anti-Tuberculosis treatment from our TU on 2001 when I was suffering with Uterine Tuberculosis. After completing 6 months' of treatment I have been feeling okay. I got married in 1990; my husband is a school teacher. I expressed all my problems and treatment to him. After listening to the truth, he has started to avoid me. I have a 10-year-old son. After starting my symptoms of discharging irregularly, he had no complain to me to sleep in the same bed, but after diagnosis of Tuberculosis, he has started avoiding me and from those days I am living in a separate room under the same floor and we have no conjugal relationship. Even to carry on our family life, he always maintains a distance.

This is all a poor women's history. RNTCP stopped my irregular discharge and lower abdomen pain by the Cat-I regimen but totally hampered my conjugal life.

Case-3: Migrant businessman

I am Mr. S.S., a tiger prawn seed businessman of 37 years age, from Sagar Island. I am doing my business since 1990. I collect the tiger prawn seeds from several fishermen and sell them in the North 24 Parganas district of West Bengal. For some time, I stayed at Nandigram of Medinipur district of West Bengal, which is my father's land, and I do my business there too. I was suffering from excessive cough and fever for long two months in 2001. I came to Rudranagar Rural Hospital and started with the medicine for Tuberculosis after my sputum examination. But after 2 months I was forced to go out from my village to Nandigram for one month and stopped the medicines. I returned and met the sister of our sub-centre when she requested me to take medicine. Again I went to North 24 Parganas for one month and felt the same problems as they were 10 months back. So I again contacted the STS. He sent me for sputum test and again started a medicine with injection. He told me to stay at home for about 8 months. Unfortunately, after completing the injections and 2 months' medicines, I left home for 3 months and stopped medicines. After 3 months, when I returned, I was again attacked with the same symptoms and I have lost my hope to live. My daughter and wife carried me to the hospital and met with the STS and doctors. They signed a bond. I also signed it to stay at home for eight months. They again started my medicines with injection. My daughter and wife forced me to stay at home. Sisters of sub-centre and STS, STLS played a very friendly role and I took the medicines regularly. After 9 months I got cured. Presently I am working well with full strength and am earning a handsome amount regularly. Thanks to all the hospital staff who came to my help and cured me. One request: is it possible to complete the medicines within two or three months which may help cure all patients like me who are forced to go out from their home to carry on the livelihood?

CONCLUSION

Rudranagar Tuberculosis Unit is isolated from the mainland. The isolated lands are almost free from pollution. Above all, the island communities are dependent on agriculture and fishing, which are their main occupations. The households of the villages are sparsely distributed. So the chances of contact are very low. Slum-like settlement patterns are almost rare. Maximum affected persons have a history of contact of TB from outside of their island, which was explained by 90% of the interviews, focus group discussions and case-studies. Medical officers also have the same view as chances of contacts with NSP are low, so the symptoms of TB suspects are not very common in the hospitals. As a result, case findings are very poor.

On the other hand, fear of identification of 'TB rogir ghar' (a house having TB patients) in the society sometimes make them aloof from the hospital. Though this person is very rare (only 3% of the in-depth interview and 2% focus group discussions supported this view). One important view has come out from the group interview of medical officers that X-ray facilities at Rudranagar TU, i.e., Sagar Rural Hospital is the cause of negative pulmonary cases. They also exhibited that extra pulmonary TB diagnosis from this remote setting is very difficult; the tests for EP TB are costly and patients have to travel to Kolkata for tests like FNAC (fine needle aspiration cytology), IgG (Immunoglobulin G), IgM (Immunoglobulin M) test, etc.

Besides the above explanation, this community is not properly aware about TB and all other diseases. They used to go to hospitals or doctors when they were not able to do anything. Their concept of diseases is absolute illness and weakness and being unable to do any work. So, it is another problem to find out the patients at the very earlier stage. Fear of side-effects and long duration of treatment are the main causes of default, though in case of NSP the default rate is within 4% to 5%, which is under the target rate of WHO guideline. The last important cause is that STS and STLS are always over-pressured to case findings which are not in their hand. As those staffs are contractual, they always suffer from fear of loss of job. As a result, there may be a chance to make falsification of report.

RECOMMENDATION BY COMMUNITY

1. Daily dose with injection is better in everyday life.
2. Duration of treatment within 2 to 3 months is better in their island economic life.
3. Panchayat members and village *morols* (headman) should be made aware about the dangers of TB.
4. Fisherman's leader (*majhi*) should be made aware about TB.
5. Continuous radio advertisement before and after the news in local radio channels is appropriate for the campaign of Tuberculosis and its vulnerability.

NOTES AND REFERENCES

1. TB India. 2002. 'RNTCP Status Report'. Central TB Division, Directorate General of Health Service, New Delhi.

2. *Monorama Year Book.* 1007. Kottayam, Kerala: Malayala Monorama.
3. Park, J.E. and K. Park. 1990. *Essentials of Community Health Nursing*. Jabalpur, Madhyapradesh: Banarsidas Bhanot, pp. 194-195.
4. Module for Senior Treatment Supervisors (STS), Part-1 & 2, Central TB Division, Directorate General of Health Service, New Delhi.
5. WHO. 1997. *Treatment of Tuberculosis: guidelines for National Tuberculosis Programmes.* 2nd edn. Geneva: WHO.
6. Park, K. 1997. *Preventive and Social Medicine.* Jabalpur, MP: M/s. Banarsidas Bhanot, pp. 138-152.
7. Bhattacharya, P. 2000. 'Current Recommendation and Chemotherapy and Pharmacology of Anti Tuberculosis Drugs'. *Journal of Indian Medical Association* 98(3): 119-22.
8. Jain, N.K., K.K. Chopra and G. Prasad. 1992. 'Initial and acquired isoniazid and refampicin resistance to Mycobacterium Tuberculosis and its implications for treatment'. *Indian Journal of Tuberculosis* 39: 121-4.
9. Das, R.K. 2000. 'Tuberculosis – Historical Landmarks'. *Journal of the Indian Medical Association* 98(3): 112-114.

23

EPIDEMIOLOGICAL INVESTIGATION OF HEALTH RISKS FROM FLUORIDE: A CASE STUDY OF TUMKUR DISTRICT, KARNATAKA

P.N. Venugopal

1. INTRODUCTION

Dental fluorosis occurs because of the excessive intake of fluoride either through fluoride in the water supply, naturally occurring or added to it; or through other sources. The damage in tooth development is from the overexposure to fluoride. Teeth are generally composed of hydroxyapatite and carbonated hydroxyapatite. When fluoride is present, fluorapatite is created. Excessive fluoride can cause yellowing of teeth, white spots and pitting or mottling of enamel. When the concentration of fluoride exceeds, 1.00 mg/litre, it induces development diseases of teeth in the calcification stage of children, which is known as Dental Fluorosis.

In Karnataka State out of 27 districts, 18 districts are affected by fluorosis. More than 200 villages in Madhugiri and Pavagada Taluk of Tumkur District are considered to be worst affected areas and 90% of the school-going children in these villages are suffering from Dental Fluorosis. Though the disease has been prevalent since a quarter of a century, no efforts have been made to investigate the epidemiology of the disease in this area. Hence, the present investigation was undertaken to know the severity of Dental Fluorosis in Tumkur district of Karnataka.

2. MATERIAL AND METHODS

Study Area: The study was conducted in Madhugiri Taluk and Pavagada Taluk of Tumkur district, where the incidence of fluorosis was reported endemic.

Selection of sample: A pilot survey of the taluk was undertaken to know the severity of Dental Fluorosis. In total, five villages were purposively selected for the study.

The samples were drawn through purposive sample method, making a total of 461 subjects aged 8-62 years. Information on socio-economic factors, food habit and nutritional status in the village was collected using pre-tested questionnaire. The observation was made with the help of Dean 's classification (1942) for Dental Fluorosis in identifying normal, questionable, very mild, mild, moderate and severe cases.

The role of water and nutrition in affected Dental Fluorosis cases were observed, recorded, tabulated, percentage calculated and analyzed using appropriate statistical tools.

3. RESULTS

Table Nos. 1-6 present frequencies and percentages of Dental Fluorosis with reference to villages and various demographic parameters.

3.1. Dental Fluorosis and Gender

53.7% males and 46.3% females were affected out of 461 observations. Dental Fluorosis is comparatively more prevalent in males than in females, though the P value of .998 shows non-significance, i.e., the fluorosis affects both the genders equally (Table No. 1).

3.2. Dental Fluorosis and Age

The maximum age group affected due to Dental Fluorosis was 8-18 years in all the 5 villages. After this age, gradual decline was observed. Mottled both enamel is restricted to the age group as fluorine gets deposited in teeth during the periods of classification of the permanent teeth. Dental Fluorosis was not present up to the age of six years. Hence, 1 to 7 age group subjects were not examined for the present study and also contingency coefficient of 0.134 and P value of .023 was found to be statistically significant for the age group. From the distribution, it is evident that in Gujjanadu 8-18 years subjects were more affected, in Vaderahalli, subjects of 19-28 years were more affected, and in Siddapura, subjects between 30-40 years were more affected than subjects of other age groups (Table No. 2).

3.3. Dental Fluorosis and Socio-economic Status

The samples were divided into four income groups as lower, lower middle, upper middle and upper class, based on the yearly income of each family

Table No. 1
Distribution of Dental Fluorosis Patients according to Gender-wise

Villages	Gender		Total
	Male	Female	
Gujjanadu	48 53.30%	90 46.70%	90 100%
Kathirajanahalli	66 52.80%	125 47.20%	125 100%
Vaderahalli	41 53.90%	76 46.10%	76 100%
Siddapura	36 53.30%	64 46.70%	64 100%
J. Achammanahalli	57 54.20%	105 45.80%	105 100%
Total	248 53.70%	213 46.30%	461 100%

Contingency Coefficient value = 0.017; P<. 998 (Non-significant).

Table No. 2
Distribution of Dental Fluorosis Patients according to Different Age Groups

Villages	Age groups in years					Total
	8-18	19-29	30-40	41-51	52-62	
Gujjanadu	39 43.30%	27 30.00%	13 14.40%	7 7.70%	4 4.40%	90 100%
Kathirajanahalli	39 31.20%	37 29.60%	25 20.00%	16 12.80%	8 6.40%	125 100%
Vaderahalli	24 31.60%	28 36.80%	11 14.50%	9 11.80%	4 5.30%	76 100%
Siddapura	22 33.30%	19 28.80%	15 22.70%	6 9.10%	3 4.20%	65 100%
J. Achammanahalli	37 35.20%	32 30.50%	21 20.00%	9 8.50%	6 5.80%	105 100%
Total	161 34.90%	143 31.1% I	85 18.40%	47 10.20%	25 5.20%	461 100%

Contingency Coefficient value= 0.134; P<. 023 (Significant).

from all source. From the observation it was found that the maximum number of Dental Fluorosis patients was found comparatively in poor class and very least number of patients was from upper class and few patients were from upper middle class. The obtained contingency coefficient found to be statistically non-significant. In other words, Dental Fluorosis has affected equally all the economic groups (Table No. 3).

3.4. Dental Fluorosis and Occupational Status

From Table No. 4 it is clear that in all five villages, the maximum number of Dental Fluorosis affected patients were students, followed by labourers and agriculturists and the minimum incidence was with housewives, businessmen and officials. Contingency coefficient value of 0.337 was found to be significant at .000 level, where we find significant association of occupations with Dental Fluorosis (Table No. 4).

3.5. Dental Fluorosis and Dentifrices

Comparatively higher percentage of Dental Fluorosis was observed with patients using toothpaste, i.e., 51% of patients used toothpaste and 26% and 22% patients were using charcoal or mud and tooth powder, respectively (Table No. 5).

3.6. Dental Fluorosis and Dean's Classification

From all the above observations, in Vaderahalli, it was observed that fluoride level in drinking water is 3.5 mg/litre. Hence, the severity of Dental Fluorosis is very high with 51.3%. J. Achammanahalli succeeds Vaderahalli and takes the second position; the fluoride level in drinking water is 3.4 mg/litre. Therefore, the severity of Dental Fluorosis is 49.5%. Further, Gujjanadu takes third position in fluoride level, that is, 3.2 mg/litre; hence severity of Dental Fluorosis is 43.3%. Compared to the above three villages, the fluoride level is less in Kathirajanahalli, i.e., 2.5 mg/litre. Hence, the maximum number of Dental Fluorosis was in moderate stage. Compared to above, all the four villages, the fluoride level is still less in Siddapura, hence the maximum number of Dental Fluorosis patients was in mild stage (Table No. 6).

4. CONCLUSION

It can be concluded from the present study that Madhugiri and Pavagada Taluks of Tumkur district containing high fluoride in its drinking water is an endemic taluk. The people consume lower amount of protective food. Age, sex, socio-economic status, occupation and use of dentifrice are some

Table No. 3
Distribution of Dental Fluorosis Patients according to Different Socio-economic Status

Villages	Socio-economic Status				Total
	Poor class	Lower middle class	Upper middle class	Upper class	
Gujjanadu	41 45.60%	27 30.00%	19 21.10%	3 3.30%	90 100%
Kathirajanahalli	73 58.40%	41 32.80%	11 8.80%	0 0.00%	125 100%
Vaderahalli	38 50.00%	26 34.20%	12 15.80%	0 0.00%	76 100%
Siddapura	31 47.60%	18 27.70%	12 18.50%	4 6.20%	65 100%
J. Achammanahalli	48 45.00%	33 31.40%	18 17.70%	6 5.70%	105 100%
Total	231 50.10%	145 3 1. 7%	72 15.60%	13 2.80%	461 100%

Contingency Coefficient value = 0.207; P< .057 (Non-significant).

Table No. 4
Distribution of Dental Fluorosis Patients according to Different Occupation

Villages	Occupational Status						Total
	Agri-culturist	Labours	Students	House wives	Business-men	Officials	
Gujjanadu	22 24.40%	28 31.10%	29 32.20%	4 4.4%	5 5.50%	2 2.20%	90 100%
Kathirajanahalli	26 20.80%	34 27.20%	56 44.80%	6 4.80%	3 2.40%	0 0.00%	125 100%
Vaderahalli	18 23.70%	21 27.60%	29 38.20%	3 3.90%	5 6.90%	0 0.00%	76 100%
Siddapura	8 12.30%	12 18.50%	20 30.80%	3 4.60%	16 24.40%	6 9.20%	65 100%
J. Achammanahalli	27 25.00%	30 28.60%	34 32.40%	4 3.80%	5 4.80%	5 4.80%	105 100%
Total	101 21. 9%	125 47.10%	168 36.40%	20 4.30%	34 7.30%	13 2.80%	461 100%

Contingency Coefficient value = 0.337; P< .000 (Highly Significant).

Table No. 5
Distribution of Dental Fluorosis Patients according to use of Different Dentifrices

Villages	Dentifrices			Total
	Charcoal + Mud	Tooth Paste	Tooth Powder	
Gujjanadu	20 22.20%	48 53.30%	22 24.40%	90 100%
Kathirajanahalli	42 33.60%	56 44.80%	27 21.60%	125 100%
Vaderahalli	13 17.10%	47 61.80%	16 21.10%	76 100%
Siddapura	18 27.70%	33 50.80%	14 21.50%	65 100%
J. Achammanahalli	28 26.70%	53 50.50%	24 22.90%	105 100%
Total	121 26.20%	237 51.40%	103 22.30%	461 100%

Contingency Coefficient value = 0.136; P< .037 (Significant).

Table No. 6
Distribution of Severity of Dental Fluorosis Patients according to Dean's Classification

Villages	Fluoride Level in drinking water in mg/litre	Dean's Classification					Total
		Quest-ionable	Very mild	Mild	Mode-rate	Severe	
Gujjanadu	3.2	8 8.90%	9 10.00%	13 14.40%	21 23.30%	39 43.30%	90 100%
Kathirajanahalli	2.5	15 12.00%	18 14.40%	24 19.20%	42 34.40%	26 20.00%	125 100%
Vaderahalli	3.5	4 5.30%	4 5.30%	8 10.50%	21 27.60%	39 51.30%	76 100%
Siddapura	1.6	10 15.40%	13 20.00%	18 27.70%	15 23.10%	9 13.80%	65 100%
J. Achammanahalli	3.4	11 10.50%	7 6.70%	13 12.40%	22 21.00%	52 49.50%	105 100%
Total		48 10.40%	51 11.10%	76 16.50%	121 26.20%	165 35.80%	461 100%

Contingency Coefficient value = 0.332; P<. 000 (Highly Significant).

of the factors influencing prevalence of fluorosis. Lower prevalence of Dental Fluorosis in female might be due to the cause of females' immigration to the study areas. Prior to their marriage, they probably were not exposed to high fluoride during their childhood; and also in males heavier physical activities which lead to higher BMR, forces them for increased consumption of water containing fluoride. Fluorosis cannot occur once the tooth has erupted into the oral cavity. At this point, fluorapatite is beneficial because it is more resistant to dissolution by acids (demineralization). This may be the reason that in our results we observe a decreasing trend of fluorosis as age increases. Although it is usually the permanent teeth, which are affected, occasionally the primary teeth may also be involved. Even though non-significant result is found in socio-economic status, it has comparatively differential effect for different economic groups. It might be deficient or inadequate due to dietary intake, which could be one of the reasons. The prevalence of Dental Fluorosis has got significant influence on students, labourers and agriculturists because of higher physical activities by them, followed by increased consumption of water containing fluoride compared to housewives, businessmen and officials, who can afford to buy expensive processed water. Our results can be comparable with the findings of Horowitz (1999), who reported that use of toothpaste containing 100% 0-1500 ppm of fluoride increases fluorosis. Finally, the observation made in all the five villages indicated that severity of Dental Fluorosis is directly proportional to concentration of fluoride level in drinking water.

BIBLIOGRAPHY

Dean, H.T. 1942. 'The investigation physiological effect by the epidemiological method', in F.R. Moulton (ed.), *Fluorine and Dental Health*. Pub. No. 19. Washigton: American Association for the Advancement of Science, pp. 23-31.

Horowitz, M.S. 1999. 'Proper use of fluoride products in fluoridated communities'. *Lancet* 353: 1462.

24

HEALTH CARE SERVICES BY UNQUALIFIED RURAL MEDICAL PRACTITIONERS: FINDINGS OF A SURVEY IN ANDHRA PRADESH

B.V. Sharma

Researches based on large scale household surveys in India pointed out that the private providers render treatment and other services related to all major public health programmes (NSS Survey 52nd Round; National Family Health Surveys). The importance of private health care providers in India in meeting the health demands of the people has also been pointed out by many others (Uplekar, 2003; Bhat, 1999; Nandraj, 1994; Nandraj and Duggal 1996; Phadke *et al.*, 1995; Uplekar *et al.*, 1998; Yesudian, 1994). However, the provision of health care in private sector is not by a single category of providers. The heterogeneity of the group is because it comprises of members formally trained in different systems of medicine, the unqualified 'professionals'[1] and the various types of ethnomedical practitioners; members rendering services through solo practice, nursing homes and corporate hospitals, etc.; members offering services free of cost and those who charge for their services. Overall, the private health sector in India is characterized by a mixed ownership pattern, many types of providers, and different systems of medicine (Nandaraj, 2001). The community of unqualified 'professionals' – more popularly referred as Registered Medical Practitioners[2] in many parts of the country – has attracted the attention of scholars interested in health studies recently for more than one reason. One issue is estimation of the strength of these providers as there is no record of them with government departments. The others are like: What are the services rendered? Who are the clients? Why are their services preferred? The researchers have broadly concluded that they are the major providers and largest sources of curative health care in our country as in the case of some of the neighbouring countries (Aljunid, 1995; Hanson, 1998; Peters *et al.*, 2004; Peters and Kayne, 2003; Bustreo *et al.*, 2003; Hosain, 2005). Similarly, it is observed that many of them give allopathic treatment (Chuttani *et al.*, 1973; Bhatia *et al.*, 1975) and their services are preferred in the rural areas (Phadke *et al.*, 1995; Nandaraj and Duggal, 1996; Uplekar *et al.*, 1998; Bhat, 1999). The studies by Hotchikiss *et al.* (2000), Jagadish and Cleland (2001), Mathew *et al.* (2001), George (2003), Verma *et al.* (2003), Iyer (2005) and others also reveled different dimensions of the utilization of the services of the unqualified providers, especially in rural areas. The implications of the findings of these studies for health policy are made clear. What is more important is that these studies also indicate the requirement of studies in other parts of the country and among different communities for national policy, owing to the cultural and other diversity of this vast country. Further, studies covering reasonable sample of providers of this community and paying attention to many issues of practice and utilization will be very valuable. This paper is an outcome of such a study in the state of Andhra Pradesh.

The study is conducted in two districts of Andhra Pradesh, namely, Krishna and Mahaboobnagar. While the Krishna district in coastal Andhra Pradesh represents the economically prosperous region of the state, the other district in Telangana represents the backward region of the state. This selection is based on the assessment of the districts on the basis of some 26 variables chosen to indicate the degree of the development or backwardness.

Brief profiles of the sampled districts:

Table No. 1
Profile of Sampled Districts

	Mahaboobnagar	**Krishna**
Total Area	18,432 sq. km	8,727 sq. km
No. of *Mandals* (Sub-district administrative units)	64	48
No. of villages	1537	955
No. of towns	11	19
Total population	3,077,050	3,698,833
Rural population	2,734,858	2,373,879
Urban population	342,192	1,324,954
Density of population	167 per sq. km	424 per sq. km
Sex ratio	973	969

Literacy	29.58	53.16
Male literacy	40.80	60.55
Female literacy	18.03	45.54
Percentage of STs	7.39	2.48
Percentage of SCs	17.63	16.58

A total of 247 and 332 RMPs in Krishna and Mahaboobnagar districts of the State were selected as samples for the study. These numbers are arrived after choosing to interview one RMP in each village in 25% of randomly selected villages from amongst the listed villages in the census records (1991). For the purpose of random sampling, the first number was chosen by the method of lottery. The first RMP contacted in that village was the sample unit. The data have been collected through interviews administering a semi-structured interview schedule during the months of December 2002 and January 2003.

I. SOCIO-DEMOGRAPHIC FEATURES OF RMPs

A) Distribution by Age

While about one-third of the RMPs in these districts belonged to the age group of less than 30 years, another 37.2% are in the age group of 30-40 years. These data suggest that the number of those who have chosen the profession of RMP has actually increased during the last 10 years. The relatively young age of a significant number of RMPs also suggests that many of them commenced the practice at an early age.

Table No. 2
Distribution by Age Groups

Distribution by age groups	**Mahaboobnagar**		**Krishna**		**Total**	
Age groups (in years)	**Number**	**%**	**Number**	**%**	**Number**	**%**
<=20	2	0.60	0	0.0	2	0.3
20 – 30	128	38.21	76	31.0	204	35.2
30 – 40	118	35.22	98	40.0	216	37.2
40 – 50	62	18.51	39	15.9	101	17.4
50 – 60	18	5.30	19	7.8	37	6.4
Above 60	7	2.09	13	5.3	20	3.4
Total	335	100	245	100	580	100

b) Distribution by Sex

Except for 15 out of the total of 580 respondents, all RMPs interviewed were males. Some women who have retired as public health workers and those who have taken training in the MPHW course, but have not found employment, have taken to this profession recently. The non-preference of women to this profession is largely because of the requirement of mobility in this profession, particularly when they have to depend on home visits.

c) Distribution by Caste

Table No. 3
Distribution by Caste

Caste	Mahaboobnagar		Krishna		Total	
	Number	%	Number	%	Number	%
SC	15	4.5	52	21.2	67	11.6
ST	6	1.8	2	0.8	8	1.4
BC	175	52.2	100	40.8	275	47.4
OC	139	41.5	91	37.1	230	39.7
Total	335	100	245	100	580	100

The *dalits*[3] among the RMPs accounted for nearly 60% of the total respondents. However, amongst them the majority (78%) belonged to caste groups, which are listed as Other Backward Castes (OBC) by the state government. This is perhaps because the members of these communities traditionally served as health providers, rendering such health services as massages, baby bathing, etc. Moreover, members of the OBC communities enjoy higher acceptability than the Scheduled Castes (SC) and Scheduled Tribes (ST) by upper caste communities. The RMPs belonging to Scheduled Caste community constitute 21% in Krishna district while they are only about 5% in Mahaboobnagar. This is in spite of a slightly greater proportion of SCs in the district. The lower proportion of SCs in Mahaboobnagar is perhaps due to their lower levels of educational attainments.

D) Distribution by Educational Status

Table No. 4
Distribution by Educational Status

Education	Mahaboobnagar		Krishna		Total	
	Number	%	Number	%	Number	%
Primary	3	0.9	1	0.4	4	0.7
Middle	8	2.4	120	49.0	128	22.1
High	110	32.8	107	43.7	217	37.4
College	199	59.4	9	3.7	208	35.9
University	15	4.5	8	3.3	23	4.0
Total	335	100	245	100	580	100

Except for four RMPs in the sample, all had education beyond primary level. Almost equal number of RMPs reported to have been educated up to High School or college in the two districts. While about 60% of the RMPs in Mahaboobnagar are educated up to college, those with similar educational qualification in Krishna accounted for only about 4%. Thus, a great majority in Krishna are high school drop-outs. The slightly higher proportion of SCs among the RMPs in Krishna contributes to the variation observed with regard to educational qualifications of the RMPs in two districts. Similarly, it may also be true that those educated up to college or beyond in Krishna have greater opportunity for some other employment; but this is not so in Mahaboobnagar because of its backwardness.

II. FORMAL TRAINING AND/OR WORK EXPERIENCE

A) Formal Training, if any and the System of Medicine in which Trained

About 50% of the RMPs studied some course of relevance for their practice and also obtained a certificate. However, such course in majority cases, was not a full-time professional course and the certificate issued was not by any recognized Board or a University recognized by the State. Very often, the certificate was in recognition of successful completion of training in first aid by John Ambulance or by the local branch of State Level RMP Association. A little above 90% among those who possessed a certificate claimed to have undergone formal training in allopathic system of medicine. Some obtained 'valid' certificate of 'Registered Medical Practitioner' after having undergone a course in some Indian systems of medicine. Less than 10% pursued such courses as Multi-Purpose Health Worker (MPHW), Sanitation Inspector, Health and Nutrition and Medical Lab Technician.

B) Duration of Training and the Degree of Award of Certificate

The duration of training was more than a year in the case of 42.4% in Krishna and 23.8% in Mahaboobnagar district. This is more often because it was not a full-time course, but a part-time one, the classes having been scheduled only during weekends. Those who have attended a crash course and completed the course in about three months comprised one-fourth of the total respondents.

C) Apprenticeship/Work Experience and System of Medicine Trained in

Except for a meager 6%, all respondents have learned the skills and knowledge for their practice while working with an allopathic practitioner. Almost 84% of them had worked under a private qualified allopathic practitioner. About 10% of the respondents gained the relevant experience while working under another RMP and very often their *gurus* were their kinsmen and the profession was inherited. The RMPs who reported to have undergone training under another RMP were more in Mahaboobnagar than in Krishna district.

Table No. 5
Qualification of the Doctor under Whom Worked

Qualification	Mahaboobnagar		Krishna		Total	
	Number	%	Number	%	Number	%
MBBS	248	74.0	180	73.5	428	73.8
MD	14	4.2	18	7.3	32	5.5
MS	7	2.1	28	11.4	35	6.0
RMP	43	12.8	11	4.9	54	9.3
BAMS[4]	11	3.3	1	0.4	12	2.1
BHMS[5]	00	00	1	0.4	1	0.2
DHMS[6]	00	00	1	0.4	1	0.2
NR	12	3.6	5	2.0	17	2.9
Total	335	100	245	100	580	100

As a large percentage of respondents had worked under a qualified allopathic practitioner, they had training in allopathic system of medicine. Even those who worked with other RMPs or under those qualified in other systems of medicine also claimed to have been trained in allopathic system.

D) Duration of Apprenticeship

Table No. 6
Duration of Apprenticeship in Years

Duration	Mahaboobnagar		Krishna		Total	
	Number	%	Number	%	Number	%
<=3	133	39.7	39	15.9	172	29.7
4 – 6	114	34	76	31.0	190	32.8
7 – 9	44	13	48	19.6	92	15.9
Above 9	32	9.6	50	20.4	82	14.1
NR	12	3.6	32	13.1	44	7.6
Total	335	100	245	100	580	100

The duration of apprenticeship varied from less than three years to more than nine years, but more often it was 4-6 years in the case of RMPs in Krishna district and less than three years in Mahaboobnagar. Those with more than seven years of work experience constituted about one-fourth in Mahaboobnagar, but RMPs with similar level of experience accounted for about 40% in Krishna district. Thus, compared to RMPs in Mahaboobnagar, those in Krishna district had less educational qualifications, but greater work experience.

III. PRACTICE

A) Number of Years since the Commencement of Private Practice

The RMPs in Krishna district possessed relatively greater experience in private practice. The average number of years in private practice in the case of RMPs in Krishna district worked out to be 12.49 years, while for the RMPs in Mahaboobnagar, it is 11.06 years. The number of years since the commencement of private practice was less than six years for about one-third of the RMPs. This also suggests that many have got into this profession during the last 10 to 12 years. The RMPs with an experience of private practice of more than 18 years constituted only about one-fifth in these two districts.

B) Mode of Practice

The mode of practice clearly varied in these districts. The fixed based providers among the RMPs in Mahaboobnagar constituted the near majority (49%). More than two-thirds in Krishna, however, provided services through home visits while also having a static clinic. The relatively greater presence of qualified providers and also unqualified providers in Krishna district, as the respondents asserted, resulted in 'stiff' competition. The RMPs, as they say, require maintaining a good rapport with their clients and home visits is one strategy in this regard. The institutions of annual payments and observance of caste hierarchies also lead to rendering of services through home visits. The spread of hamlets in larger geographical areas and also the poorer health awareness among the inhabitants of Mahaboobnagar district, perhaps discourage RMPs to opt for rendering services through home visits.

c) Villages Covered in Case of Home Visits

RMPs in both the districts covered up to a maximum of 10 villages. However, 57.6% of the respondents restricted their home visits to just one or two villages. The RMPs who covered less than four villages accounted for 90% of them. Since the population of many villages in Mahaboobnagar is too small, there are more RMPs in this district covering five or more villages than in Krishna district.

Table No. 7
Villages Covered in Case of Home Visits

Villages	Mahaboobnagar		Krishna		Total	
	Number	%	Number	%	Number	%
1	53	31.0	61	31.3	114	31.1
2	41	24.0	56	28.7	97	26.5
3	27	15.8	35	17.9	62	16.9
4	25	14.6	28	14.4	53	14.5
5	9	5.3	4	2.1	13	3.6
6	9	5.3	6	3.1	15	4.1
Above 6	7	4.1	5	2.6	12	3.3
Total	171	100	195	100	366	100

IV. SERVICES DELIVERED

a) Provision of in-patient care

A very significant percentage of RMPs reported to be rendering in-patient (I.P) care. Further, such facility is more often extended by the RMPs in Mahaboobnagar than by those in Krishna. This may be attributed to the demand of such services owing to the inaccessibility of the inhabitants of Mahaboobnagar to qualified providers both in public and private sector.

Table No. 8
Provision of IP Care

IP Care	Mahaboobnagar		Krishna		Total	
	Number	%	Number	%	Number	%
Yes	83	25.3	24	11.2	107	19.7
No	245	74.7	191	88.8	436	80.3
Total	328	100	215	100	543	100

B) AVERAGE PATIENT LOAD

Three-fourths of the RMPs had a patient load of more than 10 a day. About 20% in the two districts also treated 30 or more patients a day. More in Krishna than in Mahaboobnagar have reported a patient load of 30 plus a day. The home visits by the RMPs as well as the higher educational attainments and economic status of the members in Krishna may be the reason for quick response to any morbidity and hence greater patient load for the RMPs.

Table No. 9
Average Patient Load per Day

Patient Load	Mahaboobnagar		Krishna		Total	
	Number	%	Number	%	Number	%
<=10	79	23.6	62	25.3	141	24.3
11- 20	132	39.4	90	36.7	222	38.3
21-30	73	21.8	37	15.1	110	19.0
30 +	51	15.2	56	22.9	107	18.4
Total	335	100	245	100	580	100

Children constituted less than 25% of the total patient load in case of 58% of the RMPs. RMPs in whose case children formed the major (50% +) category accounted for just one per cent. The majority of them treated equal number of males and females on an average. Thus, the majority of the RMPs do not carry any special reputation as specialists of male, female or children from the patient perspective. This is true in case of both the districts.

c) Main types of services delivered

As many as 18.3% of the RMPs were offering in-patient services along with out-patient services. Double the proportion of RMPs in Mahaboobnagar (24%) than in Krishna (11%) extended this service. A little more than 55% of the RMPs accepted to be undertaking minor operations such as sutures, incision and drainage (opsis), fracture reduction, and POP casing, DNC, uterus cleaning, cervictomy, etc.

Table No. 10
Main Types of Services Delivered

Services	Mahaboobnagar		Krishna		Total	
	Yes %	No %	Yes %	No %	Yes %	No %
IP care	23.9	76.1	10.6	89.4	18.3	81.7
Minor operations	58.5	41.5	35.5	46.5	56.4	43.6
Deliveries	65.7	34.3	49.8	50.2	59.0	41.0
Abortions	11.3	88.7	4.5	95.5	8.4	91.6
Diagnostic lab facilities	7.5	92.5	11.8	82.2	9.3	90.7

It is interesting to note that though a great majority of the RMPs are men, 60% of them reported to be rendering their services for delivery. RMPs rendering this service are clearly more in Mahaboobnagar, as they constituted 66% in this district, as against 50% in Krishna. The data suggest that gender of the health provider is no barrier as far as this service is concerned. Those conducting abortions as part of their health services in Mahaboobnagar accounted for 11.3%, as against 4.5% in Krishna. The implications of these services by unqualified providers is noted in many studies (Varkey, P. *et al.*, 2000; Jain *et al.*, 2004). The great majority of the RMPs (90%) did not have a pathological lab attached to their clinics and this is much more so in case of RMPs in Mahaboobnagar.

d) Preference or claim of specialization

A slightly more than one-fifth of the RMPs in the two districts claimed some specialization. As stated earlier, the proportion of children among their patients did not exceed 50% in case of more than 90% of the RMPs. But half of the RMPs reported to be specialists of children among the 127 who claimed some specialization. Similarly, 44.4% in Krishna and 30.5% in Mahaboobnagar claimed to be specialists of 'females'.

Table No. 11
Claim of Specialization for Illnesses of Specific Sex or Age Group

Specialization	Mahaboobnagar		Krishna		Total	
	Number	%	Number	%	Number	%
Males	11	13.4	13	28.9	24	18.9
Females	25	30.5	20	4.4	45	35.4
Children	46	56.1	12	26.7	58	45.7
Total	82	100	45	100	127	100

Table No. 12
Illness of Specialization

Specialization	Mahaboobnagar		Krishna		Total	
	Number	%	Number	%	Number	%
Epilepsy	3	4.9	0	0.0	3	3.1
TB	10	16.4	4	10.8	14	14.3
Leprosy	1	1.6	1	2.7	2	2.0
Piles	4	6.6	0	0.0	4	4.1
Gynic	5	8.2	8	21.6	13	13.3
Paralysis	3	4.9	1	2.7	4	4.1
Delivery	5	8.2	2	5.4	7	7.1
Jaundice	20	32.8	13	35.1	33	33.7
Mental	1	1.6	0	0.0	1	1.0
Asthma	2	3.3	0	0.0	2	2.0
STDs	1	1.6	0	0.0	1	1.0
Others	6	9.8	8	21.6	14	14.3
Total	61	100	37	100	98	100

About 17% of the respondents claimed specialization for one or a group of illness. The majority (33.7%) claimed specialization for jaundice, followed by TB (14.3%) and gynic (13.3%). More RMPs in Krishna than in Mahaboobnagar claimed specialization for gynic problems. This claim of specialization of RMPs was generally based on under whom they have acquired their skills and knowledge. For example, when they worked under pediatrician, they claimed specialization on child diseases. As many of the RMPs worked in nursing homes or maternity clinics in Krishna, the proportion of those claiming specialization for gynic problems is also relatively higher in Krishna district. The range of illnesses for which specialization is claimed is higher in Mahaboobnagar than in Krishna. Thus, some RMPs in Mahaboobnagar claimed specialization for diseases like epilepsy, piles, mental illness, asthma, STDs, etc. along with the others mentioned earlier.

E) Mode of treatment

Though nearly 60% RMPs resorted to supplying drugs directly, 94% also stated to be prescribing drugs. More in Mahaboobnagar (74%) resorted to direct drug supply than in Krishna (35%). Thus, the RMPs supply drugs which are not expensive and insist on purchase of the drugs from the counters when they are expensive. Those who give drugs directly also claim that they seek payment for the drugs only and not for consultancy. The RMPs reported routine administration of B-complex injections to gain patient satisfaction. They would also prescribe and administer injections 'as per the need'.

Table No. 13
Mode of Treatment

Mode	Mahaboobnagar		Krishna		Total	
	Yes %	No %	Yes %	No %	Yes %	No %
Prescription of drugs	91.0	9.0	98.4	1.6	94.1	5.9
Administration of injections	97.3	2.7	91.0	90.	94.7	5.3
Supply of drugs directly	74.0	26.0	34.7	65.3	57.4	42.6

Except about 10% of the RMPs, all referred their patients for some diagnostic tests, 'as and when needed'. Most often, the tests were related to blood and urine. About 10% of the RMPs referred their patients for sputum examination for diagnosis of TB along with others and a small percentage advised tests for HIV.

F) Extension of preventive and promotive health care services

Half of the RMPs claimed to be offering preventive services in some way or the other. About 70% claimed extending services of immunization to pregnant women and children. RMPs in Krishna more often extended immunization to children than in Mahaboobnagar. Similarly, 55% reported giving necessary services for delayed pregnancy and family planning.

Seventy to 90% of the RMPs offered promotive health care services like nutrition education, on matters of public health, treatment facilities in public health facilities, and health risks for malaria, chest diseases and diarrhoea. The number of RMPs informing about health insurance are, however, few and constituted only about 25%.

Table No. 14
Nature of Health Advice

Services	Mahaboobnagar		Krishna		Total	
	Yes %	No %	Yes %	No %	Yes %	No %
Nutrition	92.1	7.9	90.4	9.6	91.4	8.6
Spread of diseases	94.0	6.0	86.2	13.8	90.6	9.4
Facilities in govt. hospitals	78.9	21.1	61.5	38.5	71.4	28.6
Medical health insurances	24.9	75.1	26.4	73.6	25.5	74.5
On habits and heart diseases	67.5	32.5	65.7	34.3	66.7	33.3
About cause and prevention of Malaria	86.4	13.6	72.0	28.0	80.2	19.8
About cause and prevention of diarrhoea	81.1	18.9	67.8	32.2	75.4	24.6

V. REFERRAL PRACTICES

A) Circumstances under which referred

Except a very negligible number, all the RMPs in the two districts reported referring their patients under certain circumstances to other qualified providers in private and public sector. They asserted referring patients very often during emergencies such as accidents with multiple fractures, poisoning, etc. or when, in their opinion, some major step is essential. However, more RMPs in Mahaboobnagar (93.6%) than in Krishna (78.7%) reported referring patients in emergencies to others. Eighty per cent of the RMPs also stated to refer patients in difficult deliveries. But again some significant differences are observed between the two districts in this regard. The deliveries were generally conceived to be difficult when the BP of the pregnant woman is very high, when there is intensive bleeding, when the lady experiences 'very prolonged labour', or when breech presentation is observed.

Table No. 15
Circumstances of Referrals

Circumstances	Mahaboobnagar		Krishna		Total	
	Yes %	No %	Yes %	No %	Yes %	No %
In emergency cases	93.6	6.4	78.7	21.3	87.4	12.6
Difficult deliveries	86.3	13.7	69.8	30.2	79.4	20.6
Treatment failure	75.0	25.0	59.6	40.4	68.6	31.4

Specialist treatment	86.0	14.0	84.7	15.3	85.4	14.6
On patient's preference	65.5	34.5	76.6	23.4	70.2	29.8
When patient cannot afford	59.8	40.2	46.0	54.0	54.0	46.0
Other reasons	14.6	85.4	11.5	88.5	13.3	86.7

The RMPs generally were reluctant to accept any situations of treatment failures as they also say they would accept to treat only for 'minor illnesses' and in case of 'emergencies' or any suspicion of 'major illness' they would any way advice their patients to consult other qualified providers immediately. However, 59.6% in Krishna and 75% in Mahaboobnagar stated to be referring their patients to others when their patients do not experience expected relief within a specific time period.

As many RMPs explicitly or implicitly claim sufficient knowledge to judge whether the illness is minor or major, they also expressed referring patients to qualified providers for specialist treatment in case of 'major' illness sooner or latter. Interestingly, less number of RMPs in both the districts claimed that they refer patients to others on patients' preference. The practice of referral in cases when the patient could not afford treatment was stated by even less number of RMPs. As many as 46% of RMPs denied referral of patients to others in view of the patient's inability to afford for treatment.

B) Preferred choice in case of referral

The RMPs claimed referral to providers in private sector, both generalists and specialists and also to public hospitals. However, referral to other RMPs is denied by all except 20 in two districts taken together. Three-fourths of the respondents asserted to be referring to public facility as and when needed. However, such instances occurred for specific chronic diseases like TB, leprosy when the patient's economic status is poor or because they were convinced that the treatment facilities in public hospitals are better for such diseases. Referral to public facility is more in case of RMPs in Mahaboobnagar (80.8%) than in Krishna (65.5%). This may be due to the proximity of the district to the state capital and availability of reputed institutions for the needed medical care. The RMPs in Krishna district generally preferred to refer to individual practitioners than to hospitals or nursing homes. The qualified-unqualified networks appear to be more clearly defined and firmly established in Krishna District than in Mahaboobnagar. These linkages are often through clearly defined terms and conditions of mutual benefits.

Table No. 16
Practitioners to Whom Referred

To whom would you refer	Mahaboobnagar		Krishna		Total	
	Yes %	No %	Yes %	No %	Yes %	No %
Private qualified general practitioner	64.3	35.7	56.2	43.8	60.9	39.1
Private qualified specialist provider	64.3	35.7	66.0	34.0	65.0	35.0
Referral to private hospitals	61.6	38.4	44.7	55.3	54.5	45.5
Referral to govt. hospitals	8.8	19.2	65.5	34.5	74.4	25.6
Referrals to other PMPS	2.7	97.3	4.7	95.3	3.6	96.4

Though three-fourth of the RMPs asserted referring patients to public facility, 38.7% also stated that they have not referred even a single case to public facility during a reference period of three months. This inconsistency is clearly observed in Krishna than in Mahaboobnagar. The RMPs who have referred one or two cases during the reference period account for 57% in Mahaboobnagar, while they constitute 39% in Krishna district.

The RMPs asserted that they resort to referral to other private qualified or public facility immediately on their understanding that it requires specialist treatment. Thus, 44.9% of the respondents stated that they have referred their patients during the reference period without any treatment by them. About three-fourths referred within two days of commencement of treatment by them. However, about 20% in Mahaboobnagar stated to have treated their patients for seven or more days prior to referring to other qualified providers.

Table No. 17
Average Number of Days Treated Prior to Referral

Average number of days reated by the provider	Mahaboobnagar		Krishna		Total	
	Number	%	Number	%	Number	%
Nil	45	23.7	114	16.5	159	45.9
One	43	22.6	18	11.0	61	17.2
Two	26	13.7	14	8.5	40	11.3
Three	19	10.0	9	5.5	28	7.9
Four	11	5.8	3	1.8	14	4.0
Five	6	3.2	2	1.2	8	2.3
Six	0	0.0	0	0.0	0	0.0

Seven	13	6.8	2	1.2	15	4.2
Eight	3	1.6	1	0.6	4	1.1
Nine	0	0.0	0	0.0	0	0.0
Ten	1	0.5	0	0.0	1	0.3
Above ten	23	12.1	1	0.6	24	6.8
Total	190	100	164	100	354	100

When the RMPs chose to refer to a public facility, they preferred to send the patients to a referral hospital at the sub-divisional headquarters or at the district headquarters. Thus, 85.9% of RMPs in Mahaboobnagar and 93.2% in Krishna referred their patients to such hospitals only during the reference period.

Table No. 18
Nature of Institution in Case of Referral to Public Facility

Referral institution	**Mahaboobnagar**		**Krishna**		**Total**	
	Number	**%**	**Number**	**%**	**Number**	**%**
PHC	22	14.1	5	6.8	27	11.8
GH	134	85.9	68	93.2	202	88.2
Total	156	100	73	100	229	100

Only few (9.8%) RMPs stated that they do get patients referred by other qualified and unqualified providers. However, such incidents are observed to be rare. About 51% of the RMPs, who get such referrals, stated to have treated just one patient during a reference period of three months. Interestingly, 31.6% of the respondents stated that in such cases the patients have travelled more than 10 km.

Table No. 19
Referrals Received from Other Providers

Referral	**Mahaboobnagar**		**Krishna**		**Total**	
	Number	**%**	**Number**	**%**	**Number**	**%**
Yes	40	11.9	17	6.9	57	9.8
No	295	88.1	228	93.1	523	90.2
Total	335	100	245	100	580	100

Table No. 20
Average Number of Referrals Received

Average number referred	Mahaboobnagar		Krishna		Total	
	Number	%	Number	%	Number	%
One	20	50	9	52.9	29	50.9
Two	14	35	5	29.4	19	33.3
Three	5	12.5	3	17.6	8	14.0
Four	1	2.5	0	0.0	1	1.8
Total	40	100	17	100	57	100

VI. PAYMENTS

A) CONSULTANCY FEES

A majority (87%) RMPs reported charging less than Rs. 10/- towards consultancy fees. As many as a quarter of them also stated to be collecting only about Rs. 5/- towards fees. Generally, this fee included the cost of the drugs supplied directly to the patient or for administering injections. Further, it is also revealed that they do not insist on any payment if the drugs are not supplied from their stock or if an injection is not administered.

Table No. 21
Consultancy Fees Collected

Consultancy Fees (in Rs.)	Mahaboobnagar		Krishna		Total	
	Number	%	Number	%	Number	%
Five or less	100	29.9	47	19.2	147	25.3
6 – 10	204	60.9	156	63.7	360	62.1
11 – 15	17	5.1	14	5.7	31	5.3
16 – 20	13	3.9	9	3.7	22	3.8
Above 20	1	0.3	19	7.8	20	3.4
Total	335	100.0	245	100.0	580	100.0

B) ACCEPTANCE OF CREDIT AND PAYMENT IN KIND

As many as 95% RMPs stated extending the facility of credit to their clients. The RMPs in Krishna more often reported of having an agreement for annual payments with certain families in the area of their operation. Under this agreement, the RMP is obliged to render required medical services for the illnesses that he can treat to all family members throughout the year for payment in cash or kind. These payments are made after the harvest during the months of March and April. However, this agreement does not cover the cost of medicines. This practice is to some extent an extension of the institution of *jajmani* system that prevails in rural India.

Many RMPs in both districts also asserted that instant cash payments are only an exception rather than a well-accepted convention. The offer of credit facilities, as many revealed, is the key factor for the continuation of cordial relationships between the RMPs and their clients.

DISCUSSION OF FINDINGS AND CONCLUSIONS

It is evident that the 'profession' of an RMP by and large attracted the attention of the youth in the recent past. In India, the profession of medicine is always held in high esteem and probably was prerogative of the upper castes. It is this reputation of the profession which may have resulted in more 'members' aspiring to be 'doctors'[7] in their own villages. However, members had desired to take up this profession not merely for status, but for livelihood. This essentially means the members were satisfied that there is a demand for the profession. It is also evident that the opportunities for apprenticeship under qualified medical practitioners increased in recent past. The rise in the number of qualified providers in nearby towns and cities is one definite contribution in this regard. However, it is hard to conclude that the choice of the profession in all cases is a 'preferred' choice. Surely the unemployment situation compelled to make 'forced choice' in case of significant members.

The male domination of the profession can be related to many issues. Unsuitability of the profession to women as it involves 'home visits' is related to gender constructions in case different sub-cultures of the region. But more important issue in relation to health studies in this regard is the acceptability of services of male RMPs to women in general and for reproductive care in particular. The sociological dimensions of this need to be explored in view of the reported gender preferences for gynaecological care from qualified health providers.

The sociological significance of caste is underlined in many studies in India. The influence of caste identities on practice for mutual support, referral practices in treatment of clients and for mobilizing community support and public opinion for labels that help for successful practice are dimensions involved. The client preferences for treatment from providers of specific caste identity is also an important issue to be explored.

The RMPs preference for allopathic treatment is evident, for many of them sought apprenticeship under an allopathic provider. Training under an allopathic providers for a reasonable period is not merely for skill acquisition, as in many cases training under a popular qualified physian with whom the members of the village community are familiar is necessary for legitimization of claims of an RMP.

The majority of the RMPs in Mahaboobnagar were educated beyond high school and the average training period was around three years. On the other hand, the training period was around six years in case of RMPs in Krishna district who were less educated. This desire for longer periods of training is evident in case of the less educated.

Rendering the services through home visits make RMP services 'the most accessible ones'. The flexibility offered for payments to services is another significant structural adjustment resorted to RMPs in their practice for creating demand for their services.

It is evident that RMPs render curative services to a wide range of illnesses. This calls for understanding the socio-economic consequences of the use of the services of RMPs for reproductive health and tuberculosis, as about 20% and 14% of the RMPs report to be the specialists of such services. However, this also points out the need to explore possible collaboration in rendering public health services relating to reproductive health and in the implementation of Revised National Tuberculosis Control Programme, because of the priority of these programmes in public health today.

The patterns in referral practices of RMPs suggest networks created and sustained by RMPs and the qualified practitioners for mutual benefit. The networks suggest that study of organization of medical practice by private qualified providers will be incomplete without an attempt to understand these networks.

NOTES AND REFERENCES

1. The use of the term in this paper is to refer to "Any health provider other than those who have been trained, awarded a degree and registered for practice in the allopathic system of medicine or other systems of medicine but chooses the profession of medical care with allopathic system of medicine for livelihood and so is engaged full time in this profession".
2. The abbreviated form of this, RMP is also to refer to Rural Medical Practitioner. Since the unqualified 'professionals' in the study area referred themselves as R.M.Ps in the sense of Rural Medical Practitioners, this is adopted in the context of the study.
3. Dalits refer to those belonging to Scheduled Caste, Scheduled Tribe and Backward Caste communities.

4. Bachelor of Ayurvedic Medicine and Surgery, A Degree awarded after completion of formal professional education in Indian System of Medicine (Ayurveda).
5. Bachelor of Homeopathic Medicine and Surgery, A Degree awarded after completion of formal professional education in Homeopathy.
6. Diploma in Homeopathic Medicine and Surgery.
7. The RMP's claim to be Doctors in majority cases and believe that the members among whom they practice do recognize as 'Doctors' and so treat with same respect and privileges which they would extend to the 'qualified Doctors'.

BIBLIOGRAPHY

Ahmed, S.M. 1993. 'Medical pluralism: the case of Bangladesh'. *Journal of Preventive and Social Medicine* 12: 30-5.

Aljunid, Syed. 1995. 'The role of Private Medical Practitioners and their interactions with Public Health Services in Asian Countries'. *Health Policy and Planning* 10 (4): 333-349.

Bhat. 1999. 'Characteristics of Private Medical Practice in India: A Provider Perspective'. *Health Policy Plan* 14(1): 26-37.

Bhatia, Jagdish C. and John Cleland. 2001. 'Heath Care Seeking and Expenditure by Young Indian Mothers in the Public and Private Sector'. *Health Policy and Planning* 16 (1): 55-61.

Bhatia, J.C., D. Vir, A. Timmappaya and C.S. Chuttani. 1975. 'Traditional healers and modern medicine'. *Soc. Sci. Med.* Jan. 9(1): 15-21.

Bustreo, F. Harding *et al.* 2003. 'Can Developing countries achieve adequate improvements in Child Health Outcomes without engaging the Private Sector?'. *Bulletin of World Health Organization* 81: 886-95.

Chuttani, C.S., J.C. Bhatia, D. Vir and A. Timmappaya. 1973. 'A survey of indigenous medical practitioners in rural areas of five different states in India'. *Indian J. Med. Res.* Jun. 61(6): 962-7.

Duggal, Ravi. 1996. *The Private Health Sector in India: Nature, Trends, and a Critique.* Mumbai: Centre for Enquiry into Health and Allied Themes.

Hanson, K. *et al.*, 1998. 'Private Health Care Provision in Developing Countries: A Preliminary analysis of levels and composition'. *Health Policy and Planning.* Sept. 13 (3): 195-211.

Hotchkiss, David, Barun Kanjilal, Suneeta Sharma, P.R. Sodani and Gautam Chakraborty. 2000. *Household Expenditures on Reproductive and Child Health Care Services in Udaipur, Rajasthan*. New Delhi, India: The Futures Group International, POLICY Project; and the Indian Institute of Health Management Research.

Iyer, A. 2005. *Gender, caste, class and health care access: experiences of rural households in Koppal district, Karnataka*. Trivandrum: Achutha Menon Centre for Health Science Studies, SCTIMST.

Mathews, Z., S. Mahendra, A. Kilaru and S. Ganapathy. 2001. 'Women in Waiting: Antenatal Care, Care Seeking and morbidity in rural Karnataka. Results of a Prospective Study'. *Asia Pacific Population Journal* 16(2): 11-29.

Nandraj, S. 1994. 'Beyond Law and the Lord: Quality of Private Health Care'. *Economic and Political Weekly* 29(27): 1680-85.

Nandraj, S. and R. Duggal. 1996. *Physical Standards in the Private Health Sector. A Case Study of Rural Maharashtra.* Mumbai: Centre for Enquiry into Health and Allied Themes.

Nandaraj, S., A. Khot, S. Menon and R. Brugha. 2001. 'A stakeholder approach towards accreditation'. *Health Policy and Planning*. Vol. 16, No. 2, pp. 70-79.

NSS Survey 52nd Round. National Family Health Surveys.

Peters, David H., Gita G. Mirchandani and Peter M. Hansen. 2004. 'Strategies for engaging the private sector in sexual and reproductive health: how effective are they?'. *Health Policy and Planning* 19 (Suppl. 1).

Peters, D.H., R.D. Kayne. 2003. 'Bangladesh Health Labour Market Study: Final Report'. Johns Hopkins University Project Report, Baltimore.

Phadke, Anant, Audrey Fernandes, L. Sharda, Pratibha Mane and Amar Jesani. 1995. *A Study of Supply and Use of Pharmaceuticals in the Satara District.* Mumbai: Foundation for Research in Community Health.

Ram, E.R. and B.K. Datta. 1976. 'A Study on the Utilization of Primary Health Centre and Sub-centre Health services by Rural People of Miraj Taluk, Maharastra'. *Indian Journal of Public Health.* Vol. 20, pp. 134-143.

Uplekar *et al.* 1998. 'After Health Sector Reform, Whither Lung Health?'. *International Journal of Tubercular Lung Disease* 2: 349-59.

Uplekar, M. 2003. 'Involving private healthcare providers in the delivery of TB care:global strategy'. *Tuberculosis*. Vol. 83, Issue 3, pp. 156-164.

Verma, Ravi K., Sumitra Sharma, Rajendra Singh, G. Rangaiyan and P.J. Pelto. 2003. 'Beliefs Concerning Sexual Health Problems and Treatment Seeking among Men in an India Slum community'. *Culture, Health and Sexuality.* Vol. 5, No. 3, pp. 265-276.

Yesudian, C.A.K. 1994. 'Behavior of the Private Health Sector in the Health Market of Bombay'. *Health Policy Plan* 9(1): 72-80.

25

MATERNAL AND CHILD HEALTH: PERSPECTIVE FROM THE HOS

Ganesh Chandra Mallick

INTRODUCTION

The promotion of maternal and child health has been included as one of the most important issues of the world. In developed countries, it seems in a good position but the problems of maternal death and poor health of the children are very acute in the developing as well as under-developed countries. Like other developing countries of today's world, India is not the exceptional one – here more than 9% of the conception of the mothers were damaged (NFHS-2); 140,000 maternal deaths were reported in the year 2002 by WHO (cf. SRS, 2004). Under-5-year mortality shows the worst scenario in India for the last few decades (Measham, 1999; Goyal, 1994; Anilkumar *et al.*, 2003; Visaria, 1985; Bicego *et al.*, 1996; Mason, 1997; Caldwell, 1986; Murthi *et al.*, 1995; Sachdev, 1991). Claeson *et al.* (2000) reported that in the year 1998, about 2.5 million under-5-year-olds died in India. Though the children (39.6% of the total population) and the reproductive aged women (24.4% of the total population) occupy the major portion of the Indian population but the crisis of malnutrition among reproductive aged female and children is very acute. Over 50% of reproductive aged female and over 70% of the children (aged 6-35 months) were suffering from iron-deficiency anemia (NFHS-2).

The risk is connected with child-bearing in case of women and growth development for infants and children. The child survival with safe motherhood, being one of the primary objectives, the Child Survival and Safe Motherhood Programme failed to achieve its goal to cover the 80% of children and reproductive part of the Indian population (NFHS-2). Here the percentage of the people who take antenatal, natal and post-natal care is less than the developed countries. Keeping in mind these issues, the present paper is an attempt to examine the health care practices in the antenatal, natal and post-natal stage of children and mothers among the Hos of the Mayurbhanj district of Orissa.

MATERIALS AND METHODS

Data were collected through an extensive fieldwork among the Ho population of Dighapra, Dolki, Timtima and Dhangidimuta villages of Kusumi Block of Mayurbhanj of Orissa, India. The villages were multi-ethnic in nature and the Hos were chosen because of their predominance in the villages. Altogether 253 mothers of 150 families were covered under the present study. A census schedule was prepared and filled in for this purpose. Besides census survey, observation, interviews and some case studies were done. The secondary data were collected from various Census Reports, sample surveys, books, journals and from local health centres.

RESULTS

Several empirical studies have shown that the age at marriage and age at first conception have significant effect on the reproductive life of a woman, which plays a great role along with some socio-cultural factors in case of infant and child mortality as well as maternal mortality (Das *et al.*, 2004; Bhat *et al.*, 1995; Khan *et al.*, 2006; Kumar *et al.*, 1989; WHO/UNUCEF/UNFPA, 2004; WHO, 1987). Children of very young and very old mothers have a higher risk than those children whose mothers are in the prime reproductive ages (Chavan, 1992; Chakraborty, 1987; McNay, 2003). Sandhya (1991) pointed out that the chances of child survival are lower to those mothers whose age at first childbirth is low. Table No. 1 and Fig. 1 show that the Ho mothers, who give birth their first child within 18 years of their age, their children suffered more frequently from serious problems than the children of the mothers, who give birth to their first child in their prime reproductive age.

Antenatal care refers to pregnancy-related health care provided by a doctor on a health worker, in a medical facility or at home. Antenatal care should monitor a pregnancy for signs of complications, detect and treat pre-existing and concurrent problems of pregnancy and provide advice and counselling on preventive care, diet during pregnancy, delivery care, post-

Table No. 1
Age of the Women at Time of Marriage and First Conception and Frequency of their Children Suffering from Diseases

Age Gr. of Mothers	Mother's Age at Marriage		Mother's Age at 1st Issue		Number of Children Seriously Affected	
	No.	%	No.	%	No.	%
≤ 18	45	35.16	35	27.34	15	42.86
19-20	43	33.59	24	18.75	5	20.83
21-25	33	25.78	52	40.63	9	17.31
26-30	7	5.47	14	10.94	3	21.43
30+	-	-	3	2.34	-	-
Total	128	100.0	128	100.0	32	25.00

natal care and related issues (Bhave, 1989; Chakraborty, 1987; Bhargava, 1991; Pratindhi, 1987; NFHS-2). Park and Park (1989) rightly opined in this context a minimum of four ancestral visits, in the third, sixth, eighth and ninth months of pregnancy are essential. The number of antenatal care visits to the doctor and the timing of first antenatal check-up are important for the health of the mothers and also for the outcome of the pregnancy. Ideally for normal cases, antenatal care visits required after conformation of pregnancy, should be scheduled at intervals of 4 weeks throughout the first seven months, then every two weeks until the last month and weekly thereafter (MacDonald *et al.*, 1980). A pregnant woman can have an antenatal check-up by visiting a doctor or another health professional in a medical facility, receiving a home visit from a health worker or receiving a home visit from the traditional birth attendant, *Dai*.

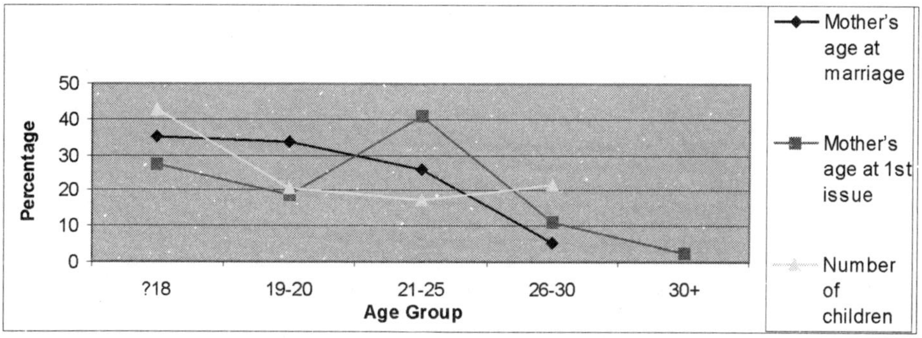

Fig. 1. Age of the women at time of marriage and first conception and frequency of their children suffering from diseases.

Table No. 2
Antenatal Check-ups of the Mothers According to Mode of Treatment

Mode of Treatment	Field area		Orissa*		India*	
	No.	%	Rural (%)	Total (%)	Rural (%)	Total (%)
Doctor	29	11.46	53.4	55.8	41.2	48.6
Nurse/Other health workers	43	17.00	25.01	23.4	18.1	16.5
Dhai (TBA)	83	32.81	0.3	0.2	0.3	0.2
No antenatal check-up	98	38.73	21.2	20.3	39.8	34.0
Total	253	100.0	99.91	99.7	99.4	99.3

Source: NFHS-2.

The present study found that in most cases, the pregnant women consulted with *Dai* during their pregnancy (32.81% of the total number of conception) [Table No. 2, Fig. 2], which is very high in comparison to state level (0.2% according to NFHS-2) and even country level (0.2% according to NFHS-2). Among them, a large proportion of mothers (38.73% of total number of conception) do not take any kind of antenatal check-up which shows a more or less similar trend with state (rural) level and country (rural) level. Among the studied population, very few women (5.93% of the total number of conception) visited doctor or any type of antenatal check-up within 3 months of their conception and a large proportion of them (29.25% of the total number of conception) took 1st antenatal check-up in the 9th month of their conception [Table No. 3, Fig. 3].

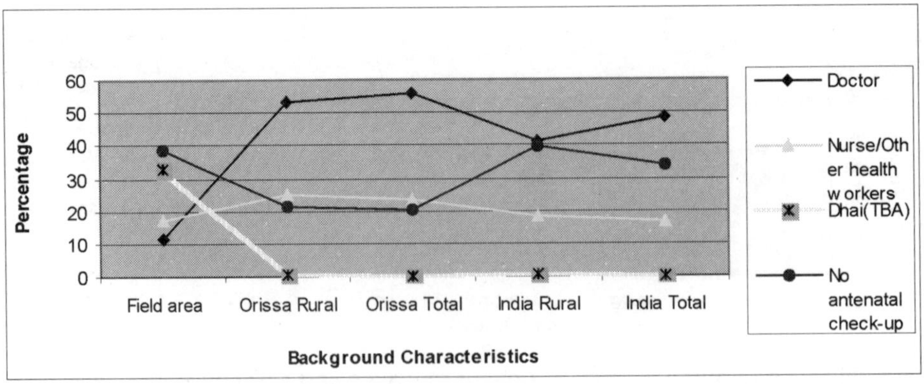

Fig. 2. Antenatal check-up of the mothers according to mode of treatment.

Table No. 3
First Antenatal Check-ups of the Mothers According to Timing of Visit

Timing of Check-up	Field Area		Orissa*		India*	
	No.	%	Rural (%)	Total (%)	Rural (%)	Total (%)
No ancestral check-up	98	38.73	21.2	20.3	39.8	34.0
0-3 months	15	5.93	10.6	10.3	8.8	8.2
4-6 months	23	9.09	22.0	21.9	14.1	13.3
7-8 months	43	17.00	23.3	22.8	14.2	14.3
9th month	74	29.25	22.7	24.4	22.4	29.5
Total	253	100.0	99.8	99.7	99.3	99.3

Source: NFHS-2.

Neonatal tetanus is most common when the delivery takes place in an unhygienic environment and unsterilized instruments are used for cutting the umbilical cord, and it is one of the major causes of infant and child death in India (Jain, 1988; Puffer, 1985; Murray, 1992). According to National Immunization Schedule, a pregnant woman should receive two dozes of tetanus toxoid injection – the first when she is 16 weeks pregnant and the second when she is 20 weeks pregnant (NFHS-2). In the study area, it is found that most of the mothers (52.96% of the total number of conception) do not take any tetanus toxoid injection [Table No. 4, Fig. 4] which shows more than double the value in comparison to state level and even in country level. Here the mothers had taken two or more tetanus toxoid injections only in case of 17% of their pregnancy whereas in state level and national level, this figure shows a moderate value (74.3% and 66.8% respectively).

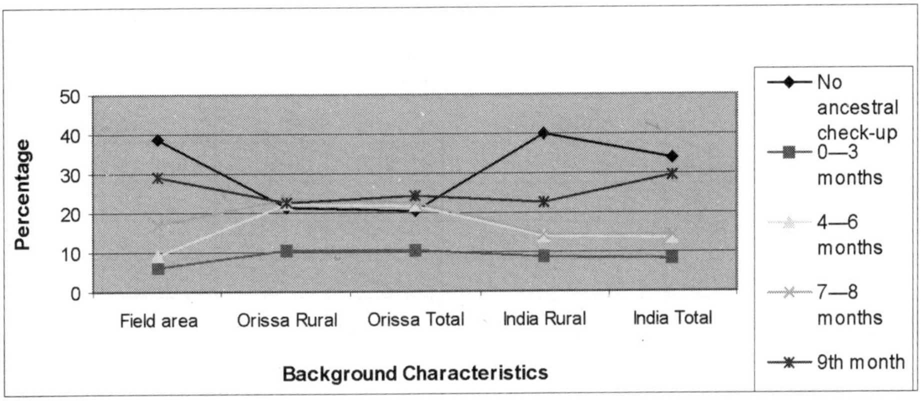

Fig. 3. Antenatal check-up of the mothers according to timing of visit.

Nutritional deficiencies in Indian women are often exacerbated during pregnancy because of the additional nutrient requirements of foetal growth (Measham, 1998). Iron-deficiency anaemia is the most common micronutrient deficiency in the world. It is a major threat to safe motherhood and to the health and survival of infants because it contributes to low birth weight, lowered resistance to infection, impaired cognitive development and decreased work capacity. Overall, about one-third of newborn children in India are of low birth weight, indicating that many pregnant women in India suffer from nutritional deficiencies. Improvement in a woman's nutritional status, coupled with proper health care during pregnancy, can substantially increase her child's birth weight. To this end, the provision of iron and folic acid (IFA) tablets to pregnant women, to prevent iron-deficiency anaemia, forms an integral part of the safe motherhood services offered as part of the MCH activities of the Family Welfare Programme (NFHS-2), now offered as part of the Reproductive and Child Health Programme. The programme recommendation is that pregnant women consume 100 tablets of iron and folic acid during pregnancy (NFHS-2). In the study area, only 26.88% of their births, mothers received iron and folic acid tablets or syrup [Table No. 4, Fig. 4], which is about half in comparison to state level and even national level.

Tetanus has long been a major killer of the newborns in India, especially in rural areas (Luther, 1998; Pandey *et al.*, 1997). Although mortality rates have fallen considerably in recent years but 2,00,000 newborns die of tetanus annually (Dastur *et al.*, 1993). The majority of the maternal deaths and much of the chronic morbidity resulting from childbirth are due to failure to get timely help for complication at delivery. The deliveries under proper

Table No. 4
Mothers Immunization Status at Antenatal Period

Name of the Vaccination		**Field Area**		**Orissa***		**India***	
		No.	**%**	**Rural (%)**	**Total (%)**	**Rural (%)**	**Total (%)**
No. of tetanus toxoid injection	None	134	52.96	18.5	18.0	28.2	24.1
	1	76	30.04	6.7	6.7	8.5	8.2
	2 & more	43	17.00	74.0	74.3	62.5	66.8
Total		253	100.0	99.2	99	99.2	99.1
Taking iron and folic tablets or syrup		68	26.88	67.4	67.6	52.5	57.6

* *Source*: NFHS-2.

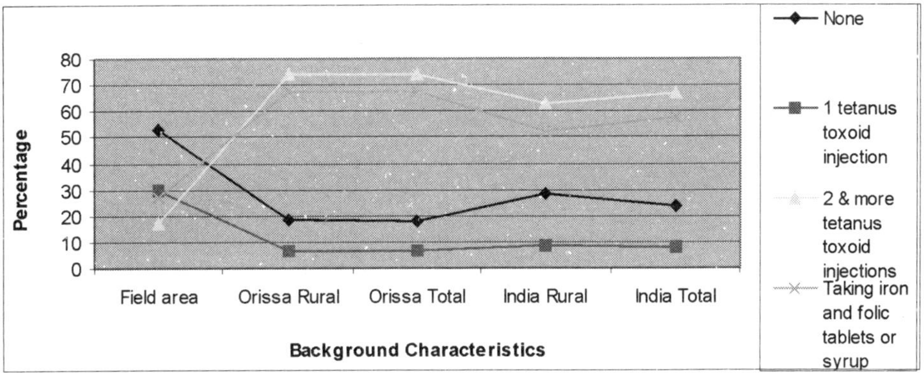

Fig. 4. Mothers immunization status at antenatal period.

hygienic conditions under supervision of trained health professional still shows a poor scenario among the Hos of the studied area in comparison to state level and national level. It has been found that in the studied area more than 88% of births took place at home [Table No. 5, Fig. 5] whereas in state level, it shows a value of 80% and in country level it shows a value of 76.7%. So, it is found that the Hos of the studied area are not aware about the modern health care system – they still believe that child is the boon of God and thus it is necessary for them that the child took birth in natural way in a purified place of their house. This conception restricts them to take modern health care facilities. Again, this conception also suggests to them to give birth a child without assistance of any person. If there are some complications, then their relatives, neighbours or traditional birth attendant (*Dai*) are called, and in very serious cases they call the doctor/ nurse or bring the patient to the hospital/private nursing home. Table No. 6 and Fig. 6 show that, among deliveries at home, 20.95% mothers have

Table No. 5
Place of Birth of the Infants

Place of Delivery	Field Area		Orissa*		India*	
	No.	%	Rural (%)	Total (%)	Rural (%)	Total (%)
Hospital	18	7.12	16.8	19.3	12.5	16.2
Privet sector	12	4.74	2.1	3.3	11.6	16.7
Home	223	88.14	80.0	76.7	74.3	65.4
Total	253	100.0	98.9	99.3	98.4	98.3

* *Source*: NFHS-2.

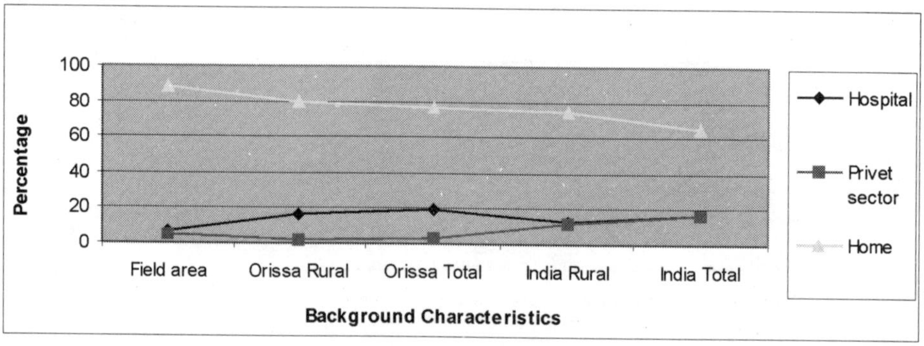

Fig. 5. Place of birth of the infants.

given birth to their children without taking help from any person, which is totally uncommon in state level and even in national level. This table also shows that the percentage of birth, attended by a doctor/nurse is very low for the studied population, which shows a relatively moderate value for state level and even country level.

The health of a mother and her newborn child depends not only on the health care she receives during her pregnancy and delivery, but also on the care she and the infant receive during the first few weeks after delivery (Desai, 1998; Kumar, 1985; Paul, 1998). Post-partum check-ups within two months after the delivery are particularly important for births that take place in non-institutional settings. Recognizing the importance of post-partum check-ups, the Reproductive and Child Health Programme (launched by the Govt. of India) recommends three postpartum visits

Table No. 6
Assistance during Delivery

Birth Attendant	Field Area		Orissa*		India*	
	No.	%	Rural (%)	Total (%)	Rural (%)	Total (%)
Doctor	18	7.11	21.5	24.3	23.0	30.0
Nurse	19	7.51	8.8	9.1	9.8	11.4
Relatives/ neighbours/ others	98	38.74	45.9	43.9	26.6	22.4
Dai (TBA)	65	25.69	23.6	22.5	39.6	35.0
None	53	20.95	-	-	-	-
Total	253	100.0	99.8	99.8	99	98.8

* *Source*: NFHS-2.

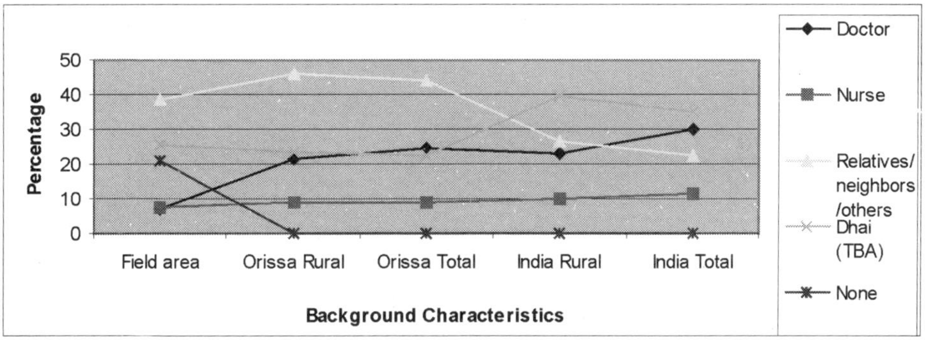

Fig. 6. Attendant at birth of the infants.

(Ministry of Health and Family Welfare, 1998b) – (i) within two days of delivery, (ii) within one week of delivery and (iii) within two months of delivery. In the studied area, only 12.25% of births were followed by a check-up within 2 days of the delivery, 25.30% within 1 week of delivery and 15.32% within 2 months of delivery [Table No. 7, Fig. 7]. So, post-partum check-up of mothers shows a similar scenario among studied population and general population of the state and nation.

The early initiation of breast-feeding is very important for the newborns because it helps them to prevent or reduce the risk of post-partum hemorrhage and facilitate expulsion of placenta. Initiation of breast-feeding starts transmission of breast milk, which contains oxytocin hormone and colostrums from mother to children, which facilitate the children's natural immunity power. In the study area, it is found that in most cases the babies (89.33% of the total births) breast-feed within 1 hour of their birth [Table

Table No. 7
Post-partum Check-ups among the Mothers

Post-partum Check-up	**Field Area**		**Orissa***		**India***	
	No.	**%**	**Rural (%)**	**Total (%)**	**Rural (%)**	**Total (%)**
Within 2 days of birth	31	12.35	12.0	19.2	14.2	14.2
Within 1 week of birth	64	25.30	28.8	11.6	30.5	30.8
Within 2 months of birth	38	15.02	19.3	28.8	16.1	16.5

* *Source*: NFHS-2.

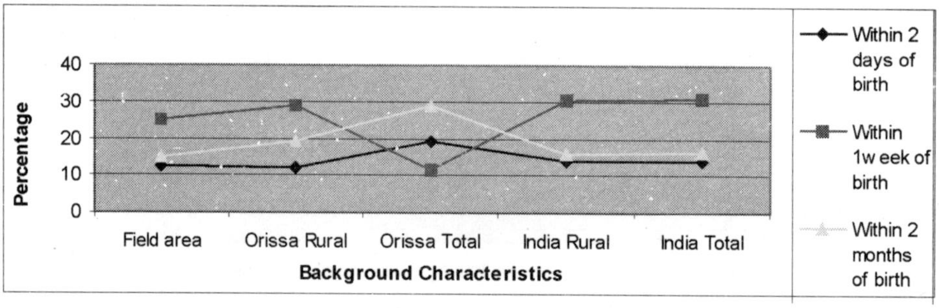

Fig. 7. Post-partum check-up among the mothers.

No. 8, Fig. 8] and few of them breast-feed within 1 day after their birth, and very few babies (0.79% of total births) breast-feed after 1 day after their birth because of some problem of their mothers during delivery. They believe that if the baby is not fed early and properly, then God will be angry. This conception influences them to early initiation of breast-feeding.

Though the Universal Immunization Programme on children against preventable childhood diseases (viz., TB, tetanus, measles, etc.) was launched by the Govt. of India to prevent childhood diseases in the country (to cover at least 85% of all infants against 6 vaccine preventable diseases) but this programme did not achieve its goal in the study area. Till now, here 32.01% of children remain outside the coverage of that programme. If we compare this picture of the study area with state level, even national level, it would be found that the studied people are far away from state level or national level. Here only 50.99% of the children had received BCG vaccination, only 34.78% of children had complete DTP vaccination and 24.11% of children had been vaccinated against measles [Table No. 9, Fig. 9].

In the studied area it was found that fever (78.26%), cough (83.40%), diarrhoea (24.90%) and skin infection (62.06%) were very common among

Table No. 8
Initiation of Breast-feeding
Among the Children

Time	No	%
Within 1 hour of birth	226	89.33
Within 1 day of birth	25	9.88
After 1 day of birth	2	0.79
Total	253	100.0

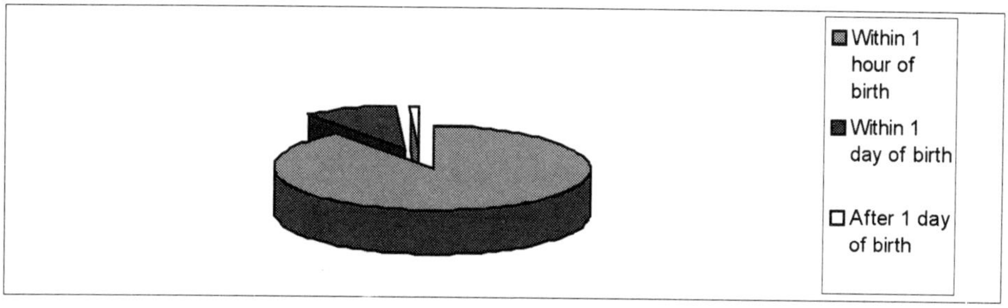

Fig. 8. Initiation of Breast-feeding among the children.

the infants [Table No. 10, Fig. 10]. Like many others indigenous communities of the world, they are not conscious about the disease. In such cases, they wait for some days for natural recovery from that problem, and if they do not get cured then they take home remedies or medicine from their traditional medicine man (*shaman*). If they do not recover by the said treatment process, they go to hospital or consult the doctor/nurse but this percentage is very low (in case of fever, skin infections only 10.10% and 1.91% of them take modern medicine respectively) as compared to the rest other treatment procedures.

FINDINGS

The Ho is one of the representatives of the most important indigenous communities of eastern India. From the above analysis, it is clear that the villagers are not all aware about the age of mothers at the time of child-

Table No. 9
Immunization Status of the Infants

Name of vaccination	**Field Area**		**Orissa***		**India***	
	No.	**%**	**Rural (%)**	**Total (%)**	**Rural (%)**	**Total (%)**
BCG	129	50.99	85.1	84.7	67.1	71.6
DTP-1	154	60.87	80.3	80.1	67.1	71.4
DTP-2	133	52.67	75.0	74.8	60.1	65.0
DTP-3	88	34.78	61.8	61.9	49.8	55.1
Measles	61	24.11	52.9	54.0	45.3	50.7
No vaccination	81	32.01	9.0	9.4	16.7	14.4

* *Source*: NFHS-2.

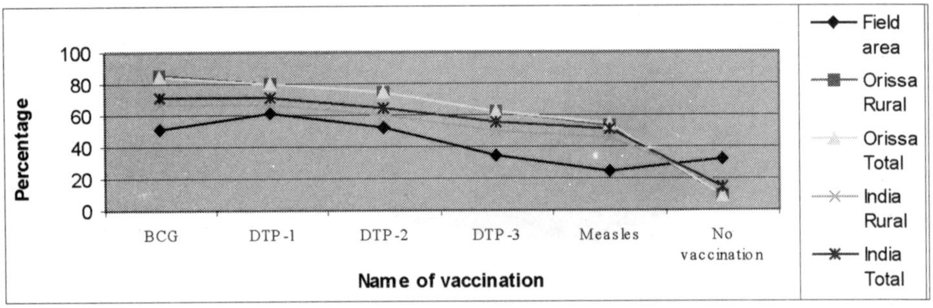

Fig. 9. Immunization status of the infants.

birth – a large proportion of them give birth to their children within the age of 18 years, which seriously affects their children's health. The study also found that till now a large proportion of the mothers (38.73% of births) have not facilitated themselves by any type of ancestral care at antenatal stage. Over 50% of the births, mothers do not take any tetanus toxoid injection or do not have iron and folic acid tablet during their pregnancy, which are very essential for forthcoming children. The reason behind not taking any injection and iron or folic acid tablets is that, they believe that the child is the boon of God, and God will be angry if they imply any external objects upon babies. They also believe that they may face problem at the time of delivery, if they fall into God's annoyance. Among them, most of the mothers give birth to their children at home, where any birth attendant is not there to assist (20.95% of births) the mothers during their delivery. Here only 15% of births, mothers were facilitated to post-partum check-ups within two months of births. It is quite encouraging that in 89.33% of births, the babies were given breast-feeding immediately after birth, which

Table No. 10
Incidence and Treatment of Diseases Among the Children

Disease			Type of treatment							
Name	Frequency		Without Treatment		Home Remedy		*Shaman*		Modern Medicine	
	No	%	No	%	No	%	No	%	No	%
Fever	198	78.26	16	8.08	85	42.93	77	38.89	20	10.10
Cough	211	83.40	48	22.75	112	53.08	42	19.90	9	4.27
Diarrhoea	63	24.90	8	12.70	26	41.27	20	31.75	9	14.28
Skin infection	157	62.06	44	28.02	83	52.87	27	17.20	3	1.91

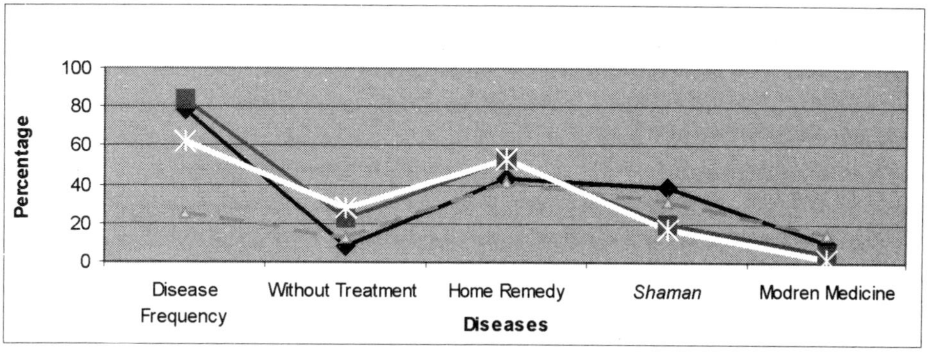

Fig. 10. Incidence and treatment of diseases among the children.

facilitates the babies to develop the natural immunity power. But in case of immunization of the children, the studied people show the same tendency as earlier – till now here 32.01% of children were far away from the coverage of Universal Immunization Programme launched by the Govt. of India, due to their communal beliefs. Even this belief influences them to recover naturally from the diseases and in case of serious problems, to take home remedy or to take medicine from the *Shaman*. As a result, higher maternal and child mortality are found in that population.

After analyzing the above said factors, the study recommends that if their communal belief against the acceptance of modern health care system would change with the help of governmental or non-governmental organization or any type of voluntary organization, they would appreciate heartily the health care programmes. This may bring a better health scenario for that population in the near future.

BIBLIOGRAPHY

Anilkumar, K. and A. Asharaf. 1993. *Mortality change in India since independence.* Paper submitted to the XVII IASP Annual Conference, Annamalai University, Annamalainagar, Tamil Nadu, 16-19 December 1993.

Bhargava, S.K. 1991. 'Perspectives in child health in India'. *Indian Pediatrics* 28 (12): 1403-1410.

Bhave, S.A. 1989. 'Trends in prenatal and neonatal mortality and morbidity in India'. *Indian Pediatrics* 26 (11): 1094-1099.

Bicego, G. and O. Ahmad. 1996. *Infant and child mortality.* Calverton: Macro International Inc.

Bhat, P.N.M., K. Navaneetham and S.I. Rajan. 1995. 'Maternal Mortality in India: Estimates from a regression model'. *Studies in Family Planning* 26: 217-232.

Caldwell, J. 1986. 'Routes to low mortality in poor countries'. *Population and Development Review* 12: 171-220.

Chakraborty, A.K. 1987. 'Neonatal mortality'. *Indian Journal of Public Health* 31 (4): 215-216.

Chavan, Y.S. 1992. 'Causes of early neonatal mortality'. *Indian Pediatrics* 29 (6): 781-783.

Claeson, Mariam, Eduard R. Bos, Tazim Mawji and Indra Pathmanathan. 2000. 'Reducing child mortality in India in the new millennium'. *Bulletin of the World Health Organization* 78: 1192-1199.

Das, R. and J. Sharma. 2004. 'Antenatal, Natal and Postnatal Health Care Practices among the Idu-Mishmis of Arunachal Pradesh'. *South Asian Anthropologist* 4(2): 169-175.

Dastur, F.D., V.P. Awatramani, S.K. Chitre and J.A. D'Sa. 1993. 'A single dose vaccine to prevent neonatal tetanus'. *Journal of the Association of Physicians of India* 41: 97-99.

Desai, S. and S. Alva. 1998. 'Maternal education and child health: is there a strong causal relationship?'. *Demography* 35: 71-81.

Goyal, R.P. 1994. 'Mortality in India: trend and prospects'. *Demography India* 23 (1&2): 103-116.

International Institute for Population Science. 2001. 'National Family Health Survey: India, 1998-99'. Bombay: International Institute for Population Science.

International Institute for Population Science. 2001. 'National Family Health Survey: West Bengal, 1998-99'. Bombay: International Institute for Population Science.

Jain, A.K. and P. Visaria. 1988. 'Infant mortality in India: an Overview', in A.K. Jain and P. Visaria (eds.), *Infant mortality in India: differentials and determinant*. New Delhi: Sage Publications.

Khan, K.S, D. Wojdyla, L. Say, A.M. Gulmezoglu and P.F. Van Look. 2006. 'WHO analysis of causes of maternal death: a systematic review', *Lancet* 367: 1066-74.

Kumar, R., A.K. Sharma, S. Barik and V. Kumar. 1989. 'Maternal mortality inquiry in a rural community of north India'. *Int J Gynaecol Obstet* 29: 313-319.

Kumar, V. and N. Datta. 1985. 'Intervention strategies for reduction of infant mortality'. *Indian Journal of Pediatrics* 52 (415): 127-132.

Luther, N.Y. 1998. 'Mother's Tetanus Immunisation is associated not only with Lower Neonatal Mortality but also with Lower Early-Childhood Mortality'. *National Family Health Survey Bulletin* 10: 1-3.

MacDonald, C. Paul and J.A. Pritchard. 1980. *Williams Obstetrics* (16th edition). New York: Appleton Century Croft.

Mason, K.A. 1997. 'Gender and demographic change. What do we know?', in R. Jeffrey and A. Basu (eds.), *Girls Schooling, Autonomy and Fertility Change in South Asia*. New Delhi: Sage Publication.

McNay, K., R. Cassen and P. Arokiasamy. 2003. 'Why are uneducated women in India using contraception? A multilevel analysis'. *Population Studies* 57: 21-40.

Measham, A. 1999. 'The performance of India and Indian states in reducing infant mortality and fertility 1975-1990'. *Economic and Political Weekly* 34 (22): 1359-1367.

Measham, A. and M. Chatterjee. 1998. *India wasting away: the crisis of malnutrition in India.* Washington: The World Bank, Health, Nutrition and Population Unit, South Asia Region, Report No. 18667.

Murray, C.J.L. and L.C. Chen. 1992. 'Understanding morbidity change'. *Population and Development Review* 18: 481-503.

Murthi, M., A.C. Guio and J. Dreze. 1995. 'Mortality, fertility, and gender bias in India: a district-level analysis'. *Population and Development Review* 21: 745-782.

Pandey, A., M.K. Choe, N.Y. Luther, D. Sahu and J. Chand. 1998. 'Infant and child mortality in India'. *National Family Health Survey Reports 11.* Mumbai: International Institute for Population Sciences and Honolulu, East-West Center.

Park, J.E. and K. Park. 1993. *Essential of Community Health Nursing.* Jabalpur: Bhanot Publishers.

Paul, V.K. and A.K. Deorari. 1998. *Newborn care in South-East Asia Region – Current status and priorities.* Report of the WHO/SEARO Regional Expert Group Meeting, Dehli, November 16-17.

Pratindhi, A. 1987. 'Infant mortality in rural India: a strategy for reduction'. *Indian Pediatrics* 24 (8): 619-625.

Puffer, R.R. 1985. *Mortality in infancy and childhood in India.* Report of the USAID, New Delhi: USAID.

Sachdev, H.P.S., P.U. Iyer and S.K. Bhargava. 1991. 'Secular trends in infant and preinatal mortality in India – implications for child survival'. *Indian Pediatrics* 28 (12): 1411-1418.

Sample Registration System. 2004. *Annual Report.* New Delhi: Office of the Registrar General, India.

Sandhya, S. 1991. *Socio-Economic and Cultural Correlates of Infant Mortality.* New Delhi: Concept Publishing Company.

Visaria, L. 1985. 'Infant mortality in India level, trends and determinants'. *Economic and Political Weekly* 20 (32): 1352-1450.

World Health Organization. 1987. *Studying Maternal Mortality in Developing Countries: A Guidebook.* WHO/FHE/87.7. Geneva: World Health Organization.

WHO/UNUCEF/UNFPA. 2004. *Maternal Mortality in 2000: Estimates Developed by WHO.* UNUCEF, UNFPA.WHO/RHR01.9. Geneva: World Health Organization.

26

DISEASE AND ITS TREATMENT AMONG THE SANTALS: AN ANTHROPOLOGICAL EXPLORATION

Premananda Panda, Swati S. Mohanty and H. Samantaraya

Among the tribal communities, the ritual is a form of language. Gabreilla states that all rituals are some form of imperfect language where linguistic procedures are followed. They have symbolic meanings (1977). This has been well established in anthropological literature. In primitive society, the ritual head, on behalf of his community, communicates to the supernatural entities for community well-being (Evans-Pritchard, 1934). An anthropologist studying such people wishes to make those beliefs and practices intelligible to himself and his readers in which a kind of rationality is demanded by the culture (Winch, 1964). Rituals are social facts. The rituals convey information about the participants and their traditions. The rituals include some orders which refer to sequences of words and actions invented prior to the current performance of the rituals in which they occur. The ritual head, on behalf of the person or community, communicates to the supernatural entity to offer sooth and console the victim and bring back the harmony in emotional order. This is how the communicator is a prop for the emotional support of the community. Religion and magic do not just explain things and help people accomplish goals. They also enter the domain of feelings. They do not have just explanatory or cognitive functions but have emotional ones as well (Kottak, 1997). The expression may occur in form of religion or magic. The magic intending to produce evil is known as black magic, in contrast to the socially and psychologically beneficent white magic. However, any form of magical task is to control, direct and manipulate the power. Similarly, it is also stated that magic and religion can reduce anxiety and allay fears. Ironically, rituals and beliefs can also create anxiety and a sense of insecurity and danger (Radcliffe-Brown, 1962). To negate the evils, the religious heads perform rites and rituals and communicate the message of the people concerned to the respective divine forces in order to harvest blessings, reduce the intensity of calamity and bring tranquil in community by negotiating the conflicting forces. Evans-Pritchard in his Nuer study mentioned such events when he referred to the role of the Leopard-skin Chief. Parallels are there in almost all communities. Community healers through rites and rituals, while acting as go-betweens, respond the desire of the sick on the one hand and the supernatural forces on the other, and negotiate between them. Dr. Krippner says that we do not have all the answers to the pressing problems of health, illness and disease. We can learn not only from western biomedical science, but from the study and observation of alternative and complementary practitioners. The study of human experience is a critically important part of science; the observation of indigenous and native practices is a part of science; the use of non-linear, systems-oriented methodologies is a part of science (1997).

Data on the role of the *Ojhas* and the rituals performed by him mentioned in this paper have been collected from the migrant Santals who settled in Village Dengabahali of District Dhenkanal, Orissa around three generations back. The history of the village refers to four families who were brought here from Mayurbhanj (a district of Orissa with more than 58% of tribal concentration) by a Chromites miner and settled in forest area as labour families. This village is numerically dominated by the Santals and only two Mahanta households, who recently came, shared their habitation in the village with due permission of the early Santal settlers.

The word *Ojha* is not of Santal origin. In Hindi language this word refers to a diviner, a sooth giver, a sorcerer, an exorcist and a magician. Many of the *mantras* used by the *Ojha* are not in Santali language. It is mostly in a corrupt form of Hindi, Oriya and Bengali mixed dialect. According to Santal culture, there are two types of spirits, the malevolent and benevolent. The rector of white magic is locally known as *Ojha* or diviner and medicine-man whereas the one of black magic is locally known as *Dan*. The *Ojha* through his skill and acumen seeks to expose and counteract the anti-social activities of the sorcerer, witch and the evil influences of the impersonal forces or powers. His magic is regarded as constructive so long as he does not inflict damage to the belongings and the health order of his fellow-man.

The lifecycle rituals in Santal society broadly refer to those rituals that confirm the stages like pregnancy, birth (*janam*), initiation (*sika*), puberty (*juin*), marriage (*bapla*), death (*maran*) and rebirth (*hunapuri-nuapuri gahr*).

Dogulas quoted the statement of Van Gennep that the danger remains in non-definable transitional stages wherein a person passes through from one stage to another (1966). The Santals believe that a pregnant woman must carefully guard herself from unsatisfied souls of the dead and the evil eyes of the living souls. She and her husband must observe certain taboos to protect the mother and the child for safe delivery. In case there is apprehension of difficulty, the *Ojha* is called upon to use his acquired divine power to identify and counter the *bonga* that does not allow the baby to come out. As per direction of the *Ojha*, a midwife, who belongs to untouchable caste, is asked to help in delivery. Soon after the child is born, the news reaches the *Ojha* and the women gathered nearer to the mother keep on beating the roof by stick to drive out the devils and ghosts moving and watching the event. The child birth pollutes the household, its members and the village geographical space and, therefore, purification of the people and the space is essential. The purification of the space is done by the *Ojha*. When the child has grown up to enter into next stage of social category through a collective initiation ceremony, locally known as *caco chhaitor*, the child is expected to shoulder the responsibility of an adult. Traditionally this ceremony is done only after the child and its contemporary belonging to an extended lineage attain puberty wherein the girls are tattooed (*khoda*) and all boys put a burnt scar on the left arm. Traditionally this is a mark signifying the societal permission for the sexual behaviour. Any deviation to this initiation is treated as misfortune. *Ojha* reminds the adults about the dreaded events likely to crop up if it is not done in time. Similar initiation process is also found among the Kharias, the Juangs and the Mundas of Sundargarh, Keonjhar and Mayurbhaj district of Orissa respectively.

In marriage, the role of *Ojha* is only a invited gentleman locally known as *bhadralok.* Only in case of repeated widowhood, the *Ojha* is called upon by the community as a whole to organize a ceremony to please all kinds of *bongas,* namely, *hapram bonga* – ancestral spirits, *jam bonga* – the death god, *huder bonga* – the spirit of recently dead and not yet became member of the ancestral spirit, *bahre bonga* and *sima bonga* – the spirits responsible to cause disease to men and animals, *jaher bonga* – the village god and *orac bonga* – the tutelary deity. The two spirits, namely, *Huder raja* and *Jam raja* are considered to be the most unpredictable *bongas* who never tolerate the prosperity of any and are treated as dreaded spirits. Contrary to this, the *jaher bonga* and the *sima bonga* are the benevolent spirits. To control the dreaded spirits, the *Ojha* cannot perform the ritual alone and, therefore, needs at the minimum two disciples who should remain in fast for the entire day and must clearly utter the *mantras* and the magical instarents as directed by the *Ojha*.

In case of sickness, the *Ojha* through his magical skill forces *bahre bonga* and *sima bonga* to go away from the house of the victim to perform the ritual to make *bongas* happy and go away from the house and the village.

The Santal rituals can be put under four headings: first, rituals reaffirm the ordered relationship between God, man, nature and the land; second, rituals seeking the blessings and cooperation of ancestors; third, rituals that seek to guard against the short-lived trouble-shooting spirits, pollution and evil eyes and the fourth, rituals that deal with the protection of potentially life-destroying circumstances and other dreaded malevolent spirits.

A Santal *Ojha* is believed to possess some kind of special powers to do things different. Through his cultivated power, he is able to perform certain amazing things that are outside the knowledge and power of his community members. When such a status is acquired, it is referring to something that first removed him from everyday contact by a 'rite of separation' which places him in the realm of the sacred in Durkheim's sense of word. The Santals believe that an *Ojha* has retained a special power attached to living and non-living objects locally known as *bongas*. The *bonga* can do and undo things when satisfied with the performance of the *Ojha*. Due to this, a social distance is maintained between the *Ojha* and his people. In Santal culture, a Santal expresses a sense of fear towards the malevolent spirit and other impersonal powers while the *Ojha* is not expected to reveal any form of trepidation. Even he is considered as a crisis manager. The *Ojha* does not express any admiration when called on to help his people to get rid of the evils and misfortunes. When he is in action, he addresses the malevolent *bongas* with abuses and bullying word. The Santals believe that their *Ojha* can crush, haft and burry the malevolent spirits. Some spirits moving in forest can cause physical damage to *Ojha* when he is under pollution. Therefore, the *Ojha* and the malevolent spirits are anti-dote to each other. Not all *bongas* are equally bestowed with power and performance. An *Ojha* after careful calculation fixes one *bonga* against another by alluring the power and knowledge of his tutelary *bongas*.

The *Ojha* not only has the power to communicate directly with the spirit, but he is also a physician having sufficient skill and knowledge of detecting the cause of ailment. He goes to the stage of transcendent and orders the spirits to behave as directed. He prescribes the treatments through the magical properties of locally available common substances and objects. By enchanting *mantras*, he professes the technique of finding out the intensity and extent of problem and deals with the causes of misfortune or frailty. This he does by taking recourse to divination maintaining that his body is possessed by a spirit who speaks through him, indicating the nature of the disease. Through the 'ultra voice' and body language the *Ojha* deciphers the desire of the malevolent spirit. While in action, he expresses messages through wide eyes, roaring sound and frequently nodded head. Frequently, he looks up to heaven with puffed mouth and tight-raised fists to the direction where the spirit abodes. During this time the people witnessing events would be talking with subdued tone all praises and power and 'success' stories of their *Ojha*. No onlooker is supposed to chide anyone during the investigation.

Santal myths have high esteem for their *Kamur Guru* – the trainer who in remote past conferred the skill of controlling power to his loveable Santal disciples. They believe that their ancestors at their native land, learnt the art of Ojhaism from a dignitary called *Kamur Guru* who is respected at par with the other tutelary *bongas*. *Kamur Guru* is invoked in the majority of rituals while the *Ojha* enchants *mantras*. As a part of shared culture, no community member has ever dared to raise any doubt to this statement.

The Ojhaship is not hereditary. It passes from one generation to another through an elaborate initiation ceremony and rigorous training in isolated places during dark night and hot sun. Not all disciples acquire the skill to acclaim *Ojha* status. Many of them remain unsuccessful and are never graced with the *bonga* power. The knowledge, thus, is scarce and therefore competition among the trainees goes deep. Anything that remains scarce and useful in society assumes ritual status. Ojhaship is one among them.

Among the Santals, the Ojhaship is androgenic. A female prophesizing warm familiarity with any type of *bonga* is looked down upon with contempt. Her presence is treated as symbol of ominous. She is viewed as inauspicious, a witch, a symbol of bad omen for the community, a body to shelter evil spirits and the like. In case of any mishaps in the village, she is attributed as a source and hence persecuted. The *Ojhas* adopt various strategies to project their importance in the community. Every *Ojha* has to establish his own reputation by demonstrating his power in public. The successful *Ojhas* are treated as superstar. An *Ojha* called on to treat patients across his community is considered very prestigious. Normally he is summoned by his fellow community members for reasons like health check-up, fixing time for ritual, calculating the marriage match, etc. The Ojhaship has its public face (normative rules) and his private wisdom (the pragmatic rules).

Unlike an ordinary Santal, an *Ojha* has a scheduled and structured way of life indicating lot of restriction on food, mobility and sex. At the out-set, an *Ojha* is to abstain from sexual activity prior to any divination or exorcism. It is believed that while he was on the way to acquire power, a set of opposite forced try to quash his virtues. Acquisition of power, through veneration to deities and spirits is to have a strong footing of the *Ojhas* between the two worlds. This is possible only with those who had surrendered themselves to the will of *Kamur Guru* and carefully maintained goodwill of the recently captured spirits during the training. He is expected not to offend even unconsciously to his tutelary *bongas* because 'He' is constantly watching him with an intention to be with those who please 'Him'.

In Santal culture, to have sex with anyone other than his wife is considered as a private sin and never a public offence. However, in case of *Ojha*, it is believed that his superstitious extra-marital sexual transaction yields him incurable diseases and reduces him to an ordinary member of the community. He becomes powerless and his acquired skill no more remains functional. He is not expected to receive any remuneration for his services rendered to others but anything that is offered to *bongas* through him is enjoyed by him. It is because it is believed that he is inseparable from the power. Normally, the household benefited through his services offers raw food materials to the *Ojha* for keep all *bongas* associated with him in good humour.

It is interesting to note here that in case of health disorder, rarely the Santals call upon the doctor. The Santals in villages remotely located with least health infrastructure facilities, like the present village, believe that if they go to the doctor, then the *bonga* will get jealous and will see to it that the illness is not cured. When the *Ojha* is called in, he first inquires for how long the patient has been ill and what type of treatment he has received till date. He then feels the pulse of the patient and looks at his tongue, eyes and presses the abdomen. The Santals believe that the *Ojha* can diagnose and cure the disease in this way. To do so, he recites various *mantras* and sings some *jharnis* (magical incantation; a type of semi-uttered sometimes subdued and sometimes in melodious voices) over the patient. As a result of their inherent magical power, these *mantras* and *jharnis* are said to have the power to drive away all evils. Through them the *Ojha* coaxes and threatens the malevolent agent. If the patient does not recover, the *Ojha* takes recourse to divination so as to ensure the ultimate cause and origin of the disease.

All societies have healers who care for the sick (Landy, 1977). Like Jhankras in Binjhal society, among the Santals their *Ojha* is he who cares for health management of his community members. Several methods of divination and exorcism used by an *Ojha* to find out the cause of sickness and to try and cure the patient, counteracting the mischief brought about by the evil spirit and impersonal powers are done. In tackling the illness, an *Ojha* shows great optimism except in obviously hopeless cases. He always exhorts the patient and his relatives not to be afraid of anything or harbour any anxieties, since he professes to have the power to cure the disease. Whenever an *Ojha* is unable to bring about a cure, so as to save his reputation, he never admits that the lack of success is due to his own inability or insufficiency of his remedies, but he remarks that the destructive spirit and power has gain the upper hand. The healer's interaction with the patient is often surrounded by a symbolic cultural concept (Kleinman, 1975).

It is difficult to say whether the *Ojha* believes in the possibilities of any revelation brought about through the various methods of divination. In any case, it is to the *Ojha's* interest to build on the Santal belief that prolonged sickness is the result of a super-normal power and, therefore, his services are required to appease these powers time and again. Very few would be there to challenge on *Ojha's* vision. It is believed that such or any form of distrust might entail disrespect to the super-natural entities, resulting in greater calamities in form of loss of cattle wealth, late monsoon, draught, famine, death of relatives and loss of property by wild animals.

Once a cure is achieved, the relatives of the patient inform the *Ojha* that they were ready to fulfil their vow and perform the sacrifice. A day is fixed and the officials of traditional institution, namely, the *Manjhi*, the *Paranik* and various relatives are invited. After taking his purificatory bath, the *Ojha* prepares the *khonda* (the cooking spot) in the courtyard and sacrifices the animal invoking the *bongas* to whom the vow had been made. The scarified animal has to be distributed as per the customary law. The head of the animal sacrificed is kept aside for the *Ojha*. The liver and lungs are cooked in rice and eaten by the *Ojha*, *Manghi*, *Paranik* and the household head. The rest is divided into three portions. One for the *Ojha*, one for the household head and the other is given to all those present. After everyone has eaten and drank *handia* (rice beer), the *Ojha* is given his fee comprised of Rs. 10/- and a low-priced cloth locally known as *dhoti*.

The Santals have their own cosmology. All calculations are done centring the *san bonga* – the sun god. The Santals recognize a regular science of Ojhaism. It is customary for young men of a village to go through an elementary course of training in this science, since it is considered for Santal men to have some knowledge of Ojhaism. In Dengabahali village and the neighbouring settlements, this course lasts for one month, starting in the end of *Bhadon* (August-September) and ending in *Aswina* (September-October) every year.

During this course, the disciples assemble every evening for about two hours in the *Ojha's* courtyard. Here they are taught various *mantras*. *bandannas* (charms) and *jharnis* to be used on different occasions. Almost all lifecycle rituals witness these. They have to learn the words and the manner of application. In the case of *jharnis,* they learned the melodies. Certain elementary medicines to be used when the *mantras, bandhanas* and *jharnis* prove ineffective are also taught. The course also comprises of various songs, dances, play-acting and method of spirit-possession *(Reem),* most of which are to be used during the *Dasaedarn.* This is only a preliminary course in Ojhaism. It does not make the participant fit to practise as an *Ojha* but only gives him an initial insight into this science. To be initiated as an *Ojha,* one has to undergo a rigorous and specialized training as an apprentice to an *Ojha.* This short course of instruction comes to an end two or three days before the *Dasaeparab* which commences during the Durga Puja. Any one who has undergone instruction and who still remembers the song, dances and method of spirit-possession can participate. The participants are locally known as *Dasae Kora.*

On the morning of the *Dasaedarn,* all those taking part accompany the *Dasae Guru* to a place on the outskirt of the village. Here the *Dasae Guru* sacrifices three fowls and two pigeons to *Kamru Guru* and the other *bongas.* The fowls are then dressed after being sprinkled with ritual water and then cooked with rice and eaten by all those present. Thereupon, the *Dase Kora* enter the village dancing and singing along the village street locally known as *kulhi.* This ritual is called *belboron.* On reaching the Manjhi's house, they enter the courtyard in single file, singing and dancing. Meanwhile, the *Dasae Guru* sprinkles some ritually purified water mixed with maize flour on the entrance door after which he is given a stool to sit on. The members of *Dasae Kora* then begin their impersonification or go for mock-play. Most of the things they personified are animals. This continues until a woman of a house has given them maize-cobs. In the same manner, they visit each house in the Village Dengabahali and also the Santal's house in the neighbouring settlements, returning in the evening to the *Dasae Guru* house, where they leave their spoils. This wandering and beginning continues for two to three days.

On the last day of the Durga Puja, all the participants accompany the *Dasae Guru,* dancing and singing all the way. They visit the Durga (Goddess Sakti) shrine installed in the northern direction of the village late in night. All the *Dasae Kora* assemble where all try to out do the other in its singing, dancing and play-acting. A week or so later, the animals which were brought are sacrificed and the flesh divided equally among all the participants. This is considered to be beneficent not only to the person so instructed but also to the village community as a whole in their fight against the evil spirit and powers. Normally if a couple cannot beget any child, they consult the *Ojha* who prescribes various medicines. If this proves to be of no effect, recourse is taken to placate supernatural forces. The women approaches an *Ojha* who identifies whether *bongas* are jamming the system. If this is so, the *bonga* is said to be transferred to a fowl and then solemnized.

The Santals believe that a birth is rarely attended with any difficulties. In case of a difficult delivery, an *Ojha* is called upon to diagnose whether a *bonga* is in attendance. If this is accepted, a vow is taken to make the proper sacrifice in the case of a speedy delivery. If the prospective mother had taken any medicine from the *Ojha* so as to become pregnant, then the child receives the name of the men from whom medicines and advice had been obtained. *Ojha* is tabooed from eating before divination. Like that the sexual act is said to be a mysterious power and is consequently tabooed before and during such auspicious occasions like remaining engaged in rituals and getting ready for hunting expedition as a member of the group.

There are a few spirits who are most personal of the Santals. Their names are kept secret and kept concealed under any circumstances to be known to other people. They are only known by the head of the household who keeps them secret till his death. Just before death, it is expected that he whispers them to his eldest son. This is very important. It seems that he can solemnize the sacrifice to the aged *bongas* only after he has learned their names. The Santals believe that even uttering the names of such *bonga* may cause them utter calamities. Special care is taken to keep this secret away from women. It is believed that if a witch gets hold of an aged *bonga,* an important member of the household will die. In such cases of warrant, an *Ojha* has to be consulted and special sacrifices must be performed as directed by the *Ojha*.

The Santal relation towards the supernatural entities is one of the reverential fear, dependence, submission and propitiation. Communication with this spirit is concretely manifested mainly through supplication, rice-beer offering and animal sacrifices that are made on behalf of a particular group, be it the whole village, sub-clan or household. To this the ceremonial sharing of sacrificial food and rice-beer libations must be added besides certain special observances and taboos. An important point to remember from the very beginning is that religion, like culture, has everywhere been undergoing a certain degree of change through a process of acculturation. Acculturation includes those processes of change which result from constant and continuous contact condition. One can distinguish two major types of acculturation: one is incorporation and another is directed change. A process which facilitates the retention of original custom as well as the acceptance of the new element is called syncretism. In case of Santal rites and rituals, there have been marked changes inclined towards the Hindu rites and rituals. Since spatial change does not lead to social change (Bailey, 1977) the traditions are retained among the Santals. In the migrated places, the Santals have been acculturated with Hindu elements. Thus, the Santals have experienced syncretism.

The *Ojha,* who is one of the most respected persons in a village, is believed to have special retained *bongas,* through whose aid the forces of evil and sorceries are countered. The Santals do acknowledge the natural causes of illness. Yet, if the disease is of unusual character and does not respond to various medicines taken, it is attributed to the displeasures of a *bonga* or to the presence of a malevolent agent. Similarly, when a number of misfortunes occur within a short time in a house, the members suspect the presence of a malevolent agent. In such cases, an *Ojha* is called upon to indicate the cause and the nature of the malady or misfortune. He then performs the *sunum-bonga* ceremony, or the divination to discover what spirit has been offended and how to appease it. In doing this, he maintains that his body is simply possessed by one of his tutelary *bonga* who speaks through him. During performance, the *Ojha's* pitch of the voice goes hand in hand with the opening of eyes and stretching of face muscles.

These special *bongas* with whom the *Ojha* is believed to be in communication and whose power he enlists in the time of need, are many and different. The *bongas* are invoked according to the prediction and faith of the particular *Ojha* called upon. Some *Ojhas* have a predilection for one, others for another. And certain *bongas* are only invoked for particular diseases, for example: *Bhalua bijaijulumpaika, ulumpika* and *kapikara* in the case of women whose lower portion of the abdomen is protracted, *lurilora* – small pox, *duria bardu* and *nasankandi* – madness, *gosae era* against sores, *jaher era* and *nanganagini* for cattle diseases.

The *bongas* who are specially remembered and invoked are *Kamuru Guru* (from whom according to Santal tradition, the *Ojha* originally learned the profession), *Sharma Sin Thakur*, sometimes also called *Sin bonga* (the sun), the *Kali mai* (the Hindu Goddess Kali), the *Dibi mai* (the Hindu Goddess Durga) and the *Ganga mai.* These *bongas* are invoked by the *Ojhas* at the opening ceremony of the training course during which the young men of the village are taught the fundamentals of *Ojha* science. Among the Santals, when ritual remains as a language of subordination and appeases to the supernatural entity for support may be referred to a religious act and when the ritual is a manipulative one to subordinate the supernatural power, we may call it as a magical act.

The role of the spirit-medium or diviner (*Ojha*) is indispensable among the Santals in which they operate and the *Ojha* is thought as one of the major social support systems and, therefore, insignia of community solidarity and hence inseparably beneficent one. The *Ojhas* have, due to their association with certain power of arresting and keeping the evil spirits away and the power of healing the pain, now a respectable status among the locals cutting across the Santal culture. In the locality at the time of uncertainty, health-hazard and misfortunes, the *Ojhas* are called upon by the non-Santals for propitiation. They often are the substitute to the doctor for the people living in close proximity with the Santals. It has been observed that in few economically sound Santal households, the Brahmins are invited to perform rituals. The doctor is called upon to treat the sick, however, they also summon the *Ojha* as and when required. Thus, we believe that any discrepancies that are being marked have not changed the Santal underlying system of beliefs on which the rituals for the *rite de passage* are founded and on which the Santals retain their identity. The Santal life and relationships between the living and non-living world is getting regulated within the framework of the religion and magical beliefs (Troisi, 1978).

All rituals demonstrate the Santal's continuity effort to cooperate with the divinities for manipulation of mystical power in order to increase or restore well being and strength at the level of household, kin group, village, and also at the larger section of the tribe. Thus, the Santal rituals can be put under few headings: first, rituals reaffirm the ordered relationship between God, man, nature and the land; second, rituals seeking the blessings and cooperation of ancestors; third, rituals that seek to guard against the short-lived trouble-shooting spirits, pollution and evil eyes and the fourth, the rituals that deals with the protection of potentially life-destroying circumstances and other dreaded malevolent spirits. All these four types appear in different form at different times during *rites de passage.*

Sarkar (1993) writes that the majority of the Santals are under the influence of the dominant caste communities (*Dikus*, meaning exploiters). Similarly, the migrant Santals of Orissa who left their native place to work in mining areas of coal, chromites and iron run by private and public sectors have their distress sale of labour and others who had no scope to engage themselves in mining activities sell their labour to the forest contractors and land-owners. Poverty often forced them to postpone their ritual and manage with non-Santal *Ojhas* at cheaper cost. They compromised some of their tradition due to economic compulsions. The process of cultural continuity and change is also reflected among the Santals of Orissa, which has opened up many assimilative processes into the Hindu peasantry and also to an extent a two-way process of tribe-caste continuum with mutual emulation of cultural traits. In this process of assimilation, superior technological, economic, political, social and cultural forces of neighbouring caste communalities affect and influence the less powerful Santal cultures, despite their cultural richness. Santal culture marginally influenced the non-Santals due to the fact that culture flows from a politically and economically dominant group to a subservient one.

The *Ojhas* are the intellectuals of their society and are the brokers of ideas between the sacred and profane world between past, present and future and between their societies and the others. They are the boundary players of many cultures because they travel to sacred spaces and spend time in trance. An *Ojha* knows the language of all *bongas* (supernatural forces) such as the *bonga* of dead ancestors and *bonga* of natural and supernatural origin. The social aspects of the *Ojhas* include all thinking and doing processes of Santal culture.

The *Ojha* is a go-between of material and non-material world. He is a negotiator of the community on whom the community keeps trust in making the evil spirits away, bringing virtues for the whole community, mending the loss incurred to the community member by positioning his power over riding the evil *bongas*. The Santal *Ojha* is structurally comparable to that of the *Jahankar* of the Binjhals of Bargarh (Panda, 1981), the *Janis* of the Kondhs of Phulbalni and *Disari* of the Bada Gadabas of Koraput (Jha, 1996). Similarly, the concept of *bonga* of the Santals, the *duma* of Gadabas and the *masans* of the Binjhals are comparable. The *Ojha* has, the *Disaris*, the *Janis* and the *Jhankars* are the persons with special power responsible for the community well-being and communicate with the supernatural entities. The dynamic relationship that exists between the *Ojha*, his comparables and the community members in tribal societies is more to be explored. Ritual as a form of communication should be analyzed with respect to the social system. We can learn more about the social meaning and function of healing. In all patients, the medicine-man encounters careful consideration of the symbolic and other non-verbal communication may be important. It will have a greater implication when we see the interaction between the *Ojha* and his clients. The role of *Ojhas* should be understood as manifestations of widespread socio-cultural phenomena rather than as something unusual or unprecedented because Ojhaism as a system of art derived from a strong shamanic tradition. It seems important to place it in a larger pan-human perspective. Such studies are expected to provide insights to understand the dynamics of change in desired direction.

In the world view of the Santals, the underlying values of the traditional Ojhaism stress that life is the product of hot and cold, and the society survives so long as there is warm air passing through the community. An individual is like cold soil who is created and shaped to become subordinate to the community because community's heating process can keep him off. The faith and surrender to power beyond the individual, therefore, are better than the individual efforts and struggle. The sources of human strength lie in a harmonious integration with one's group in the individual's affirmation of the community's value and its given order, his obedience to community's gods and goddesses and the communicator – the *Ojha* – and in his cherishing of its tradition.

The natural objects like rivers, mountains and forests have control over the productivity of land. Preparation of land for cultivation is controlled by human organization. The ancestral spirits watch the raised crops, keep evil spirits away from community and cattle wealth and informs *Ojha* on the forthcoming calamities; the village social organization is responsible for distribution of labour, land and agricultural products and organize religious activities in the community: the household activities are being controlled by the village resources and the existing social institutions in vogue, *Ojha* being its watch-dog of maintaining a balance between the tradition and the change.

The healing mechanisms of the *Ojha* operate at collective sentiments. The communication of psychic events, including mental distress through the narration of *Ojha,* is a well-known device for societal balance. The *Ojha's* suggestion to people is that it is always good to be good to people and to our neighbour and since life is composed of warm air and cold soil, the *bongas* who experienced both the worlds (the living and the dead) exist to

A tabular representation of the spirits

Malevolent Spirits

Sl. No.	**Malevolent Bonga in order of hierarchy of power**	**Place**	**Power**
	Bongas whose abodes are outskirts of the village, mountain, river and who transform themselves into snake and tiger to kill human and also create illusion.		
	Sima bongas	The village boundary spirits	(a) Believed to attack the man in field in form of poisonous snake or other cannibals (b) More malevolent than other
	Bahre bongas	Village outskirts spirits	(a) Kills the man causing accident
	Buru bongas	Mountain and hill spirits	(a) He relishes human blood
	Beghut bonga	Water spirit	(a) Creates mirage for people, especially young men

remind us the vices and virtues and we need to be selective for our lifestyle and communicate with the *bongas* for community well being as per the tradition. Thus, the Santal society is based on the principle of binary opposition.

BIBLIOGRAPHY

Abercrombie, Nicholas. 1970. 'Superstitions and Religion: The God of the Gaps', in David Martin and Michael Hill (eds.), *A Sociological Year Book of Religion in Britain*. London: SCM Press, pp. 93-129.

Bailey, F.G. 1964. *Tribe, Caste and Nation*. USA: Manchester University Press, pp. 49-51.

——. 1977. *Stratagems and Spoils*. Berkeley: Berkeley University Press.

Campbell, C. 1978. 'The Secret Religion of the Educated Classes'. *Jour. Sociological Analysis*. Vol. 39, No. 2, pp. 146-156.

Dogulas, M. 1966. *Purity and danger: an analysis of concepts of pollution and taboo*. Middlesex: Penguin Books.

Evans-Pritchard, E.E. 1934. 'Levi-Bruhl's theory of primitive mentality'. *Jour. Bulletin of the faculty of Arts*. University of Egypt.

——. 1964. *Nuer Religion*. Oxford: Clarendon Press.

——. 1956. *The Nuers: Political systems of the Nuers*. Oxford: Oxford University Press.

Gabreilla, Eichinger Ferro-luzzi. 1977. 'Ritual as language: the case of South Indian Food offerings'. *Jour. Current Anthropology*. Vol. 18, No. 3, pp. 507-5515.

Jha, M. 1996. *The Ollar Gadabas of Orissa*. New Delhi: Concept Publishing House.

Kleinman, A.M. 1975. 'The symbolic context of Chinese medicine: A comparable approach to the study of traditional medical and psychiatric forms of care in Chinese cultures'. *Jour. Am J Chi Med*. Vol. 3, pp. 103-124.

Kottak, P.C. 1997. *Anthropology: The Exploration of Human Diversity*. INC, New Delhi: Mac Graw-Hill Companies.

Krippner, S. 1977. *Stratagems and Spoils*. Berkeley: Berkeley University Press.

——. 1997. 'Medicine and the inner realities interview by Bonnie Horrigan'. *Jour. Alternative Theirapies*. Vol. 3, No. 6, pp. 98-107.

Landy, D. 1977. *Culture, disease and healing studies in medical anthropology*. New York: MacMillan.

Panda, Premananda. 1981. 'Karma: A Binjhal festival: its socio-religious significance'. *Jour. The Eastern Anthropologist*. Vol. 34, No. 2, pp. 141-150.

Rad-cliffe-Brown, A.R. 1962. *Structure and Function in Primitive Society*. New York: Free Press.

Sarkar, M.C. 1993. 'Customary Rights in Land and Forest of the Tribals in Chotanagpur – Santhal Pargana Region of Bihar', in Miri Mrinal (ed.), *Continuity and Change in Tribal society*. Bhubaneswar: Institute for Science and Development, pp. 97-108.

Shepherd, W.C. 1971. 'Religion and the Counter Culture: A New Religiosity'. *Jour. Socio. Inquiry*. Vol. 42, No. I, pp. 3-9.

Troisi, J. 1978. *Tribal Religion, Religious Beliefs and Practices among the Santals.* New Delhi: Manohar.

Victor, Turner. 1967. *The Forest of Symbols: Aspects of Nedmbu Ritual*. Cornell University Press, pp. 19-20.

Wilson, Bryan. 1976. *Contemporary Transformation of Religion*. London: Oxford University Press.

Winch, P. 1964. 'Understanding A Primitive Society'. *Jour. Arne. Anthrop. Quarterly.* Vol. 4, pp. 307-324.

27

EYE CAMPS: AN ACTION APPROACH TO CATARACT BLINDNESS

C.K. Kantharaju

India has a great tradition of voluntary service and in spite of rapid changes, the urge to serve fellowmen is still strong and widespread. Right through the ages, this tradition has been accepted as social obligation on the part of the individual towards his less fortunate brethren. Religion emphasized the virtue of *dan,* i.e., free gift, which encompassed different forms of social service. Giving cash assistance to the needy, imparting knowledge, providing food and shelter to pilgrims, care of the sick and the destitute were considered to be righteous acts. Institutions like *maths* and temples were responsible for the religiously inspired forms of service and social institutions like caste and the joint family were concerned with the care of the handicapped, the destitute, the widow and the aged. Religious considerations, by and large, motivated acts of charity and service to please God and acquire *punya.* They also generated an honest desire in a large number of people to help the needy. In the long historical process, the Indian social system developed many tendencies and practices which were by no means healthy and desirable (Lawani, 1999).

The term voluntarism is derived from the Latin word 'voluntas', which means 'will'. The will assumes various forms of impulses, passions, appetites or desires. It is prior to or superior to the intellect or reason (Muttalib, 1987).

The concept of voluntarism is based on humanism. It is based on the philosophy of treating the whole world as the 'family of man'. It has no room for difference of caste, creed, colour, race, region and religion. Rather, it is based on the principles of peaceful co-existence in spite of these disparities and differences. The term 'volunteer' is normally used to denote someone who offers unpaid service to a good cause. Man is a social animal. His gregariousness is known by the fact that he always lives in groups such as family, joint family, lineage, clan, tribe, community, etc. This is the natural corollary of social inter-dependence. Mutual aid and self-help are instruments through which a man fulfils his basic needs but can also use them for meeting or solving some of his problems. This is more important at the time of disaster and natural calamities like floods, famine, fire, epidemics, etc. People's action is, therefore, spontaneous in nature and they come together around a common interest, common need or common problem.

The eye health situation in the selected non-tribal villages and tribal settlements imperatively called for three types of action strategies; one to develop a need-based programme to create awareness about the eye health problems, the underlying causes, the available eye care services, role of community leadership, etc. on the one hand and remove misconceptions and doubts about cataract blindness, and to promote the right types of attitudes on the other; second to create a group of trained functionaries and volunteers who could undertake education as well as arrange or provide eye care services; three, to involve the health functionaries in training and provision of services at the community level. Efforts were also made to involve the NGOs through a networking arrangement.

The main interventions of action programme included:

1. Identification and enlistment of all the cataract blind individuals for screening and operation;
2. Identification of all the individuals through a check-list of signs and symptoms of cataract blindness for intensive follow-up;
3. Organizing community education programmes and motivate the cataract blind patients individually to undergo cataract removal surgery with the involvement of local leaders and local associations; and
4. Organizing eye camps through the health functionaries or referral services through Swamy Vivekananda Youth Movement Hospital, Saragoor.

The focus of the intervention strategies was on providing information on complete eye care services to all the cataract blind individuals and cataract operated patients for follow-up of post-operative complications. It included health aspects like the factors responsible for eye health problems; the importance of regular eye examination; prevention and treatment of cataract blindness and other causes of blindness.

Change in traditional beliefs and behaviours, attitudes and fear of operation can be achieved only through continuous efforts to overcome initial hesitation on the part of the people. This cannot be achieved by the health workers who visit the villages once in a while. The need, therefore, was to involve a person who clearly understood the eye health problems of people and who could act as a facilitator to solve the problems. His/her role was to identify the cataract blind patients, to prepare brief socio-economic profile of patients, to identify the barriers to low uptake of eye care services and local beliefs attached to the condition of blindness among people. Hence, the researcher in the beginning identified a number of cataract blind patients and prepared a brief socio-economic profiles of the patients in one village. After that he was trained to record the local beliefs, knowledge and attitude towards blindness in general and cataract blindness in particular. The researcher employed various techniques of data collection in the process of understanding the problems of cataract blind patients and his/her families. He also identified the impact of cataract blindness on the person such as worry, financial burden, friends' teasing, work and social problems, mobility problem, dependency problem and visual discrimination. Thus, the researcher made an effort to identify local volunteers to sort out the cataract problems. The volunteer could act as a link between the people and service providers. The main advantage of this approach was that it helped a large group of villagers who were in need of eye care services. It may be mentioned here that local leaders and local NGOs in the areas extended necessary support in organizing eye screening camps with Swamy Vivekananda Youth Movement Hospital, Saragoor.

Eye camps covered the population living within 10 km radius of the camp site. This is the usual size of the area covered by Swamy Vivekananda Youth Movement Hospital, Saragoor, in their routine eye camp service. While organizing an eye camp, local resources and resource persons like NGO, involvement of other NGOs, involvement of people, volunteers, individuals and transportation facilities were considered for identification of camp sites. The researcher voluntarily involved himself in the eye camps organized by Swamy Vivekananda Youth Movement Hospital, Saragoor. The camps were organised in association with the District Blindness Society, Mysore, Kabini Foundation, Nisana Belthur, Society for Socio-economic and Educational Development, Mysore, local leaders and respective villagers. The researcher voluntarily involved himself in the following camps during his fieldwork.

The researcher helped the camp organisers in various activities like planning the camp, publicity and in motivating a few cataract blind patients to undergo cataract removal operation at Swamy Vivekananda Youth Movement Hospital, Saragoor.

Swamy Vivekananda Youth Movement Hospital organizes eye camps in association with local sponsors, which come from a variety of sectors, e.g. business, voluntary sector or non-governmental organisation or District Blindness Control Society. Local sponsors conduct the pre-camp publicity campaign 2-3 days prior to the date of camp. In addition, they provide practical help such as arranging a suitable venue for the camp, local transportation facilities, providing food for hospital staff, etc. Eye camps are usually a one day event. Swamy Vivekananda Youth Movement Hospital ophthalmic team visits the camp site and examines the camp attendees. People who require treatment or further investigations are invited to accompany the eye camp team on hospital transport back to the base hospital on the same day.

Usually the location of an eye camp is determined by hospital personnel in consultation with the local eye camp sponsor. Swamy Vivekananda Youth Movement Hospital knows from experience that it has gained popularity in the region. The eye camps organised by this hospital are very popular among the villages, hence more publicity is not required. People cannot attend an eye camp if they do not know about it. It was necessary to standardize eye camp publicity to prevent and reduce the number of absentees. The researcher at the camp site tried to obtain general community view about the eye health problems. Camp participants were asked to identify common adult eye problems in their village, and to discuss their causes, impacts, sources of treatment and factors influencing treatment, decision making, etc. Finally, each participant was asked about the most convenient location of the camp for people in their village with cataracts to attend for eye camps and use of local hospital services.

Eye camps are conducted in partnership with a local community service organization which sponsors the camp. The major service providing hospital in the region is Swamy Vivekananda Youth Movement Hospital, Saragoor. The other agencies which are involved in arranging eye camps are District Blindness Control Society, Mysore; Society for Socio-economic and Educational Development, Mysore, Kabini Foundation, Nisana Belthur. Local leaders and respective villagers were major resource persons. After working with the sponsor in making the appropriate arrangements, the camp organiser from the hospital guides the sponsors to ensure that it adequately publicizes the date and place of camp to the surrounding community. The camp organiser visits the site about one week before the scheduled date to oversee the arrangements and to discuss with the sponsor strategies to ensure the smooth conduct of the camp.

On the camp day, the camp team, composed of ophthalmologists and paramedical staff, proceed to the camp site with support from camp sponsors and with the help of local volunteers, patient registration is done. The patient details include name, age and address in the out-patient register and case sheet is prepared. Patients are given identity cards, which may be used for any future follow-up. After this, preliminary vision testing is performed by ophthalmic assistants. Vision charts, such as the Snellen and 'E' type charts, are used. Doctors perform the preliminary examination. Clinical conditions such as external eye infections, vision loss caused by nutritional deficiency and the incurably blind are examined. Patients above the age of 45 years have their intraocular pressure tested. Trained ophthalmic assistants administer topical anaesthetic drops and measure the intraocular pressure with a Schiotz tonometer. Facilities for the patients to lie on, additional benches for waiting patients and adequate lighting are necessary. Refraction is performed on patients who exhibit refractive errors. Senior doctors evaluate the test findings, perform the final examination, review the patient records, make the final diagnoses and prescribe treatment. Patients advised for surgery are motivated by the hospital counsellor as well as by the researcher to undergo surgery at the hospital. An arrangement was also made for spectacles. Patients advised to wear glasses may purchase ready-made spectacles, if available at the hospital dispensary. Otherwise, the optician delivers finished spectacles to the patient on the next day of the camp.

Patients selected for surgery are registered in patient register and transported to the base hospital at Saragoor for surgery. These patients and accompanying attendants are provided with surgery, post-operative care, meals and transportation facilities free of cost. Later, the operated patients were advised to utilize post-operative care services and provide information on post-operative complications such as infections, etc.

In this study, the researcher was able to identify the cataract blind patients and various barriers to low of take up of eye care services. In addition, the researcher acted as a facilitator and helped a few people who were suffering from cataract blindness to regain their vision and lessen the dependence on others. He actively involved himself with service providers to arrange eye screening and surgery camps in the selected villages. He was able to initiate a dialogue with cataract blind patients and motivate them to regain their vision.

28

INDIGENOUS HEALTH PRACTICES AMONG THE KODAVAS OF KARNATAKA

S.M. Meera and C.E. Subramanya

Kodagu in Kannada and Coorg in English are derived from the local version Kodavu and the people Kodava. The name Kodava is supposed to have been derived from *Kodimalenad,* meaning higher forest land. The Puranic version says that the land of initial settlement was called Krodadesa, which later became Kodagu. The word Kodava means the people blessed by Godmother Kaveri ('Kod' means give, bless and 'avva' means mother Kaveri) is supported by the Puranic version.

The Kodavas are the major upper caste population of Coorg district. The ancestral place is called 'ainmane', the original house of the Kodavas. They speak the Kodava language. Kodava family called 'okka' is a patrilineal joint family.

Kodavas have their own special way of dressing during festivals and marriages. The male members will have a *jaripeta,* black shirt and will have a silver sword on their waist. The female members will wear beautiful and attractive sarees and dresses in their own special attractive way. Even today, the Kodavas residing in Mysore have maintained their traditional clothing on special occasions.

Coorg grows copious amounts of cardamom, pepper, rice, ginger, oranges and many other cash crops, albeit in lesser quantities. Recently, floriculture and medicinal plants growth has caught on in a big way and Coorg products have already achieved recognition as one of the best. Kodavas follow Hinduism but are more liberal and independent than any other Hindu in performing marriage, divorce, re-marriage, festivals, worship, ceremonies, dialect and dress. They worship the *Karana* (ancestors) as family deity and *Kaveri* as Godmother.

Kodava customs are also very special and attractive. City life and globalisation has not affected their traditional dressing, folk dance, folk music and folk food. They are born soldiers. This community has given great generals to our nation, much more officers and many more *jawans* to the army. This is the reason Kodavas are found everywhere and are not concentrated only in Coorg.

Important Kodagu Festivals: Kodavas follow Hindu customs in this also. Indian festivals such as Shivaratri, Ugadi, Sankramana and Shashti are celebrated. In addition, they have their own festivals like Kailapoladu, Cauvery sankramana and Huttari.

LOCAL HEALTH TRADITIONS

They date back to the cave period of man who started consuming raw vegetables and fruits when he had to satisfy his hunger and also imitate the action of the animals, which consumed plants for any health disorder.

Our traditional lifestyle is oriented towards using herbal medicines. When you showed symptoms of fever and cold, did not your grandmother give you hot pepper *rasam*? What are these practices that have become a part of our culture and evolved through many years of experience in the inhabited eco-systems? These diverse health practices are non-codified and are better known as local health traditions. Most of them are locality and community specific. This is further bifurcated into household level health traditions and *natuvaidya* traditions.

The household health tradition comprises home remedies: health related customs, seasonal regime, rituals, ceremonies food, and regimen. It not only utilises the spices and condiments from the kitchen but also a wide range of plants and its products available in the locality.

The diverse health traditions prevalent in various localities and eco-systems are better known as local health traditions. These traditions are also ethnic and community specific and knowledge is passed on by word of mouth. Local health traditions cater to the health care needs of millions of people. The carriers of these local traditions are women in ordinary village households, birth attendants, herbal healers, bone setters, *visha vaidyas* and local healers who treat domestic animals. These health practices are not documented. Field research has been undertaken in this study to assess the contemporary relevance of these practices in rural southern India.

Traditional medical knowledge of medicinal plants and their use by indigenous cultures are not only useful for conservation of cultural traditions and biodiversity but also for community health care and drug development in the present and future.

In this process, the knowledge base that has accumulated over generations also gets swept away into oblivion. This vanishing stock of knowledge is invaluable in many respects. It not only forms a locally available low input technology but also provides useful hints for future technological development. Appropriate interventions in the form of policy input, training, awareness generation, documentation, etc. can, in most cases, revitalise local traditions. Local health traditions in the Indian villages are very good examples of this. Traditional health care systems provide a spectrum of local variations such as Ayurveda, Unani, Siddha, folk health traditions, etc. Kerala has a host of these traditions still playing a vital role in the local health care sector. However, these systems have not yet been studied in-depth. For various reasons, some of these streams are on the verge of disappearance. The present work is an attempt to study the status, problems and prospects of indigenous health traditions in a selected region of Coorg. The study also emphasises on how the Kodavas have still maintained health traditions. The indigenous knowledge of health traditions and the native plants used for medicinal purposes were collected through questionnaire and personal interviews during field trips.

HOME REMEDIES AMONG THE KODAVAS

Home remedy is a set of insights in the health care acquired through experience and spread by practice. The practitioners are mainly aged women in the household. Knowledge is acquired through direct observation of the daily situations in which their elders applied it in and around the households. Home remedies are relatively simple preparations and are mostly single drug prescriptions. Here, magic and medicines act simultaneously. The combinations could be natural (only medicine), supernatural (only magic) and magico-medicinal (magic and medicine). The aged persons who have acquired special skills perform the magic. Here treatment is performed as a service without monetary compensation. Most of the home remedies are intended to solve health problems arising in the day to day life such as fever, headache, toothache, cough, eye diseases and constipation. Common ailments such as headaches or coughs are considered to be diseases with natural causes and hence their symptoms are treated at the household level, without recourse to magical practices. Some areas of Kerala, Karnataka and Tamil Nadu have reported other resources used for health care at the household level. *Keezharnelli, tulsi, adhatoda, neem,* curry leaves, coconut, drumstick, aloevera, hibiscus and castor plant are familiar names in primary health care in South Indian households.

Indigenous health practice is based on the indigenous knowledge of a given people, a given community and their experiences in the context of the local culture and environment.

Certain ritualistic acts done in the context of household also may be considered as home remedy. One such ritual is *madh toppu*. A sweet dish is prepared out of *madh toppu* (justice) herb, which is found growing luxuriantly in Coorg. This herb is believed to have 18 medicinal properties. This is harvested on the 18th day of *Karkata masa* (July 15 to August 15). The Coorgis also consume *karkata kozhi* (chicken) on this day. The reason is that during the monsoon, the chicken accumulates excessive fat content to fight the cold weather and if eaten by man, it adds to the body.

These also vary in communities and localities. The periodical cleaning of the surroundings and house, filling the house and environment with *doopas* (smoke) and many such hygienic measures are also an integral part of these traditions.

Indigenous medicine practice involves a complex combination of activities, order of knowledge, beliefs and customs to generate the desired effects for the diagnosis, prevention or elimination of imbalances in physical, psychological or social well being.

In local health tradition systems, spices, raw drugs, fruits, vegetables and weeds are also used. All parts of the plant have some use. It can be the root, stem, leaf, stamens and pistils. In the *neem* plant, the root and stem are used for fever, leaves for fever, flower for malaria, and oil from seeds to destroy lice. There are also many plants used for the same conditions – *keelanelli, manjal karisalanganni* for jaundice, *tulasi, neem*, ginger, *shaaitan*, pepper, *periayanangai* for fever.

HOME REMEDIES FOR CHILDREN

'Uppneer' is one of the remedy to treat children for stomach ache. It is basically salted water that is given to children. Besides this, the other remedy for stomach ache is ginger and lemon extract mixture called 'Shuntichorange'. 'Jaikayi Pal' is administered to the child for stomach ache and this is mixture of nutmeg (*jaikayi*) paste and breast milk (*eddhepal*). In some cases, 'Kaipakkethop' (bitter gourd leaves) ground with the breast milk is also administered to the child for the sudden relief of stomachache. 'Chundethoprasa' is also given for stomach ache. Asafoetida mixed with breast milk is also one of the remedies. Whooping cough in children is cured by administering a syrup which is made out of the extract of betel leaf, honey, onion juice and asafoetida, which is consumed for three days.

HOME REMEDIES FOR MENSTRUAL PROBLEMS

Menstrual problems were managed in the house by home remedies, which included 'omum', 'omaki' (cumin), 'nallamalu' (pepper), *neem* leaf extract, 'hing' (asafoetida), butter, 'sambhara' (coriander), jaggery, 'bollulli' (garlic) and 'gallimaddu'. Complicated cases were attended by the traditional physicians who used a large number of plants from the ambient vegetations and some products of animal or mineral origin to deal with the local diseases and ailments. But today, due to modernization and good transport facilities, they approach allopathic doctors.

HOME REMEDIES FOR SKIN AILMENTS

'Thurrikajji', a skin ailment, is cured by consuming 'payyurauche' (gomthura). This is strained four to five times and mixed with 'manjapodi'(turmeric) and consumed for one week.

'Pullukaddi' is a ringworm infection which is cured by home remedy. 'Kaipake' (bitter gourd juice), 'bevinathoppu' (*neem* leaves) and coconut oil are mixed and smeared on the body as a remedy.

REMEDY FOR WORMS AND STOMACH AILMENTS

For 'kelthalpullu' (deworming), a simple home remedy is 'papathoppu' (papaya) leaves which are eaten raw. 'Bekaipathoppu' is fed for indigestion problems. It is made from black cumin. The other remedy is 'kaipullirasa' which is the extract of bitter lemon, 'hirallekhayyi'. For loose-motion 'bekaiperassa', a syrup made from cumin mixed with water, is administered.

'Galli sonku' is the uncontrollable fever. In this 'gallimaddu' is given which is made from 'gallithoppu' (herbal leaf) and 'bolluli' (garlic) extract and is smeared on the body and then a ritual called 'akki unjith chado' is performed to ward off evil eye due to which the person is ill, by holding a handful of 'akki' (rice), 'uppu' (salt), 'malu' (redchilli) and 'kadu' (mustard). In this the body, from top to bottom, is massaged and the fever is believed to be removed. Here also, belief and medicine work complementarily. A simple scientific principle is enacted here as a supernatural phenomenon.

The other remedy to ward off fever is the herbal syrup called 'chittikodi kashaya'. This is like a syrup with the following ingredients: 'chorangirasa' (lemon juice), hot water and honey.

While, most of the common ailments were managed in the house by home remedies, which included pepper, ginger, turmeric, coriander, cumin, tamarind, fenugreek, *tulsi*, etc. Complicated cases were attended by the traditional physicians who used a large number of plants from the ambient vegetations and some products of animal or mineral origin to deal with the local diseases and ailments.

Plants such as 'doddapatre', acorous root, long pepper, touch-me-not, basil, 'datura', nuxvomica, nettle, aloe, curry leaves, emblic plants, 'bael' (bilva), 'nari visa' and 'gali maddu', as they are called in local parlance, which have been found to cure many ailments, are raised by the people. The speciality is that each herb has a distinct aroma.

The medicinal property attributed to 'gali maddu' is that it wards off evil forces from homes; 'nari visa' is grown in homes to ward off snakes and other poisonous reptiles. 'Doddapatre' juice is consumed and applied externally against measles. 'Bael' or 'bilva' is used to check obesity and long pepper is said to be a cure for cough and cold. There are herbal cures for many other forms of ailments such as herpes, allergy, kidney stones, jaundice, asthma, ulcer, and skin diseases too.

Nilamuchala (*Gymnostaychyum febrifugum*) is a febrifuge and is also used as an antidote against viper bite. Its distribution is limited to Coorg, Brahmagiris and the foothills and plains of Kannur district.

SUMMARY AND CONCLUSION

These elderly Coorg women, who are home-based healers or doctors, cater to a wide range of conditions covering promotive, preventive and curative aspects of health. They handle simple ailments like cold as well as emergencies like snakebites, traditional ophthalmology, gynaecology, paediatrics and many individual organ-related conditions and specific diseases.

This study concluded that even though the accessibility of Western medicine for simple and complicated diseases is available, some people in the studied parts of Kodagu district still continue to depend on medicinal plants, at least for the treatment of some simple diseases such as cold, cough, fever, headache, skin diseases and tooth infections.

The wealth of medicinal plant knowledge and health traditions among the Kodavas in Karnataka is based on hundreds of years of beliefs and observations. This knowledge has been transmitted orally from generation to generation. However, it seems that it is vanishing from the modern society since younger people are not interested to carry on this tradition. Due to lack of interest among the younger generation as well as their tendency to migrate to cities for lucrative jobs, wealth of knowledge in this area is disappearing. The loss of local health traditions will be a monumental loss to the world and particularly to rural communities because with the erosion of local health traditions one is losing the time-tested medical knowledge of plants, animals and minerals. It is necessary to act urgently in order to save our indigenous health traditions, which form an important facet of the Indian medical heritage.

BIBLIOGRAPHY

Abdul Hafeel, Suma T.S. 2000. *Monumental heritage, Indian health traditions.* October 08.

Busia, K. 2005. *Medical provision in Africa – Past and present*. Phytotherapy Research. 19:919-923. doi: 10.1002/ptr.1775.

Jaggi, O.P. 1973. *Folk medicine*. Delhi: Atmaram and Sons.

Madhavankutty, K. 1989. *Tribal System of Medicine: A Study of the Herbs and Medical Measure Used by Tribals of Kerala*. Thiruvananthapuram: KANFED.

Pei, S.J. 2001. 'Ethnobotanical approaches of traditional medicine studies: Some experiences from Asia'. *Pharmaceutical Biology* 39:74-79.

Unnikrishnan, E. 1995b. 'An introduction to the Ethno medicine of Kerala', in C.R. Rajagopalan (ed.), *Keraleeyatayude Nattariv* (Malayalam). Kanimangalam, Trichur.

29

ON MEDICAL ANTHROPOLOGY

R.K. Mutatkar

Anthropology as a holistic study of man has developed out of the studies of isolated people, following the ethnographic model. Health issues of tribal people have been studied by the anthropologists, which forms the bulk of medical anthropology. There has been confusion between classifying research in anthropology and ethnology. Similar confusion might appear in the areas of Medical Anthropology and Ethno-medicine.

Medical anthropology has been defined in various terms such as:

1. Bio-cultural understanding of man and his works in relation to health and medicine.
2. Encompasses the state of medical phenomena as they are influenced by social and cultural features and social and cultural phenomena, as they are illuminated by their medical aspects.
3. Medical anthropologists describe their work relating to (a) research about description and interpretation of bio-cultural interrelationships between human behaviour, health and disease levels, (b) their professional participation in health improvement programmes and programmes about changing health behaviour to promote better health.

Most of the books under the title Medical Anthropology have published papers relating to beliefs and practices of isolated communities or the folk practices of traditional peasant communities. Similarly, most of the papers presented in anthropology seminars relate to ethnographic accounts of health behaviour practices and functioning of local health practitioners. There is description, but no analysis in terms of theory or policy planning. Even the textual great health traditions of India and China, namely,

Ayurveda and Chinese medicine have been discussed under ethnomedicine in Medical Anthropology by Foster and Anderson.

Instead of discussing the practices and knowledge of shamans and herbalists under ethnomedicine, it would be more appropriate to discuss under the conceptual framework of little tradition and great tradition enunciated by Robert Redfield. The interaction between little and great tradition in India is obvious since the classical texts are generalizations of the folk practices, as are evident in home remedies and kitchen medicine. Similarly, the concept of anatomy and physiology in Indian and other traditional systems of medicine discuss metabolism and immunity. The concept of Shukra and Ojas are the guiding principles about masculinity, which hinder vasectomy to the extent of Indira Gandhi hosting political power.

Medical anthropology is applied anthropology about the health aspect of culture. It is now common knowledge that applied anthropology was rated low in anthropology. George Foster accepted that his teachers – Herskovits, Kroebar and Lowie – had successfully inoculated him against the germ of applied work, since it did not contribute to anthropological theory. Foster countered this argument by the fact that much of culture change theory came from materials collected through applied anthropology.

We are aware that after the third world countries became independent and aspired for welfare and development of the people, the expectations from intellectuals and academicians to contribute to the development process were high. In India, after the community development programme was embarked in 1952, which included the installation of primary health care in rural areas, several anthropologists from Western countries came as consultants to design extension strategies in community development. Anthropologists, and later professional social workers, were called upon to educate the illiterate people to adopt agricultural and health technologies for their development. We are aware that curative medical technologies from all systems of medicine have been accepted by the people but not those in the preventive and promotive aspects, which contradict the basic cultural norms and values. Female foeticide and age at marriage of girls do not show any change for the better. Vasectomy continues to be unacceptable. Condom is not used in marriage as safe sex tool to prevent HIV. Condom is not used widely between a female sex worker and her frequent client, with whom an emotional relationship has developed. Anthropologists understand the rationale of these behaviour patterns. We explain seemingly contradictory behaviour with the use of concepts of hard and soft core of culture. However, health education programmes having gone through the terminological change to IEC seem to be asserting on behavioural change communication to which anthropologists are supposed to contribute. The whole paradigm of health programmers is to get the people to accept what is offered to them by medical model to solve people's problems created by powerful agents in the epidemiological triangle. They want anthropologists to help them in this social marketing to accept and convince them as to how slight behavioural change will solve the problems created by bacilli and viruses.

The paradigms of health programmes are medicalised, in many cases governed by distribution of medical products coming from the pharmaceutical industry. These programmes, therefore, do not address people's concerns. For example, leprosy afflicted are concerned about social exclusion as a result of deformity, mutilation and ulcer. The programme is concerned about identifying the bacillus and killing it. Since deformity is caused by nerve damage, anthropological perspective of priority medical research should be in the area of nerve damage in preference to drug research. HIV/AIDS conveys that HIV and AIDS are synonyms, which they are not. Any counselling to HIV+ve person that AIDS could be postponed by pursuing a healthy lifestyle does not make any impact on the mind, if HIV/AIDS are communicated as synonyms.

The argument is that the medical anthropologist should address the issues of policy planning significance, using anthropological theory and research tools. The strength of anthropological research lies in qualitative methods which answer the question 'why'.

The national surveys such as NFHS, NSS and Census provide the quantitative data necessary for policy planning and budgeting. However, micro-level planning, implementation and logistics is not possible without supplementing it with qualitative data. Addressing the policy planning issues does not necessarily mean involvement in intervention. On the other hand, intervention can be transformed into intervention research which could throw leads to theory formulation. It is common knowledge for anthropologists that the interventions, if addressed to the concerns of the people, would provide meaningful outputs and would move towards sustainability by becoming a part of culture pattern. Similarly, the knowledge of integrated nature of culture, soft core changing faster than the hard core, would provide leads in appropriate planning of intervention for culture change. The strategy of 'Going Native' for establishing Rapport in field research is an exercise in confidence building for micro-planning and implementing intervention with community involvement. Mahatma Gandhi associated himself by identifying himself with the Daridranarayan by putting on their clothes and by visibly living like them. He had genuine compassion for them. We are aware that anthropologists develop an emotional attachment to tribal people they study, if they have really lived with them and indulged into participant observation. Cross-cultural research and comparative method are recommended to shed away the danger of subjectivity and inculcate objectivity.

Medical Anthropology has a bright future in India. From the one department at Pune in 1974, now all departments of Anthropology in India want to have formal courses in Medical Anthropology. It is more in demand at the ICMR and now at the AYUSH department of Government of India due to objectivity of Anthropology to plural systems of medicine and its usefulness from people's curative, preventive and promotive perspectives. Several anthropologists are employed in ICMR institutes across the country.

We should be ready to interact with the medical and bureaucratic fraternity dealing with health and development, and understand their concerns and functioning systems. Our perspective of holism and totality need not limit itself to the cultural holism of a community but must encompass all forces which are actively involved in planned change. Indigenous communities and people no longer want to remain indigenous. They want change on par with so-called mainstream society. We have to go beyond evaluation research and take on the challenges of intervention research. Ethnographic documentation has to give way to analytical research, combining the quantitative and qualitative data, although ethnography including the genealogies continue to be effective tools of data collection.

Now that the public health schools are proliferating very rapidly in India, medical anthropologists are in demand to do everything in teaching and research in the area of 'Public' in public health. Medical Anthropology does not become a social work discipline but with its theory and methodology, it has the prospect of becoming a profession.

II

Health matters have been considered as biomedical phenomena. However, people have varying beliefs and practices about personal body image, nutrition, causation of diseases, environment including plant and animal life, and fire and water. There are varying practices having a bearing on health, which relate to personal hygiene, mating patterns, child-birth and child rearing, age and sex groupings, social regulations of interpersonal and group contacts, family relationships and occupation structures. These cultural beliefs and practices and many others make health a social-cultural phenomena.

Health refers to the total fitness of an individual, which would ensure performance of social roles to his optimum efficiency. All societies equip their members, through the processes of socialization and enculturation, to play useful roles commensurate with statuses. Societies provide institutional structures to look after the sick with a view to minimize the period and extent of disability and maximize the quantum of recovery in shortest possible period. All societies provide for sick roles since they cannot afford to lose the productive and role-playing manpower through premature deaths or by falling victims to various diseases. Societies would obviously be more concerned with chronic, communicable diseases but the efforts to control these would be constrained by the knowledge about health and diseases in a society, by the capacity of the people to do organizational effort and by economic consideration.

Health is usually understood as absence of disease. In that sense, it is discussed as a medical concept giving prominence to disease control programme and medical profession. Since disease control programmes have a global perspective and linkage with the pharmaceutical industry, they get precedence over health prevention and promotion activities. India shares the global concerns of chronic, stigmatizing diseases such as leprosy, AIDS and TB. The national health concerns relate to child and women's health. In India, we have inherited plural systems of medicine such as Allopathy, AYUSH, etc. Besides, we have special issues of tribal health. Health and disease are also moral issues. A healthy person and a person who gets quick death are considered to be morally superior to those who suffer from chronic ailments and have to pray for release from suffering if they remain bed-ridden for a long time. A disease is therefore considered deviance from morality. Chronic diseases like leprosy, AIDS and TB are stigmatizing and get associated with the idea of sin, either in this birth or the past birth.

Human communities are interested in health and disease cure of its members. In the animal world, sick are left to use their own resources for recovery or die. Amongst human beings, members of the group attempt to restore the sick person to health, to enable him to perform normal roles and obligations. It is believed that in case of leprosy in the past centuries where recovery was considered impossible, the sick were left to fend for themselves. One of the methods was to invoke the divinity to help such persons, and to feed them to prevent death from starvation.

Medical systems have developed as a vast complex of knowledge, beliefs, techniques, roles, norms, values, ideologies, attitudes, customs, rituals and symbols; in short, a totality of health knowledge, beliefs, skills and practices of the members of every group. So, a medical system embraces all of the health promoting beliefs, actions and scientific knowledge and skills of the members of the group that subscribe to the system. Every society has its own concepts of health, disease causation and health care. Usually, health is understood by lay people as a totality of appetite for food, body metabolism, good sleep and body-mind vitality to perform social roles.

Conceptually, techniques and tools that provide cure or relief from symptoms are easily acceptable from any medical system as they fall into the realm of soft aspects of culture or material aspects of culture. Issues of causality of the disease or concepts of body, mind, soul, intellect and their inter-relationship with external environment fall into the realm of hardcore culture relating to the belief systems which are not easily penetrable. People can take pragmatic action for relief and cure without changing the belief system. People can accept technology without accepting science and without imbibing the scientific temper.

There can be conceptual difference between disease, a pathological concept and illness, a cultural concept. Usually, a person with disease is treated to cure his illness. The tragedy of leprosy is that a person is labelled as a patient but not as a sick person. He does not get the status of a sick person who is freed from performance of social roles. Actually, that is the malady of chronic diseases. In leprosy, a patient is not cured but only released from treatment; he never becomes a person but remains an ex-patient signifying the possibility of his becoming a patient again.

Diseases are classified according to modern biomedical concepts in terms of the transmission process, namely, communicable and non-communicable. Biomedical solutions for individual cure and public health issues are in terms of chemotherapy and vaccines to deal with causative organisms. Traditionally, medical systems have a three-fold classification: (a) curable, (b) curable with sustained effort and (c) incurable. Lay people have only two categories, curable and non-curable. They also understand the issues of mortality and morbidity. Action is taken for the symptoms which cause morbidity affecting social and economic life to get relief in order to become normal. Leprosy, although curable in biomedical sense, continues to be a problem since the biomedical paradigm and people's paradigm do not meet each other. Leprosy Elimination Campaigns (LEC) continue to target lowering prevalence with a hope of decreasing incidence in a reasonably short period. Health education, now called IEC, is aimed at operationalising this paradigm. It required considerable efforts on the part of medical anthropologists to include the third objective to leprosy control as given in the WHO document TRS 716 of 1985, to include 'prevention of associated deformities'. Similarly, the Sixth Expert Committee on WHO in

TRS 768 of 1988 modified the definition of a case of leprosy to include the patients who completed treatment and were under surveillance and those who had deformities and disabilities needing rehabilitation.

People's paradigm of leprosy control relates to freedom from fear of leprosy and not infection of leprosy. They can classify the villages and grade them according to the degree of success achieved. The indicators are level of participation in the family and community such as no divorce on account of leprosy, no denial of civic facilities and full participation in the economic activities. This freedom from fear could be achieved by campaigning about management of impairment and prevention of disabilities. This needs a constant and simultaneous dialogue with the community, family and the patients.

In the female ward of TLM Naini hospital at Allahabad, it was found that the young married women had come from their mothers' house and not from their husband's house. This data was not enquired and recorded on the case sheet. The community, patient and family have to get total knowledge about the disease process and not only the messages of curability. What is the difference between the biomedical definition of cure and people's concept of cure in leprosy? To which concept is our programme aimed at? Unless it addresses itself to the concerns of the people, it cannot meet with success. The programmers have to take initiatives in addressing its programmes to people's concerns and prepare the packages on health education to answer the question people have in mind.

To put it briefly, from the anthropological perspective, it is necessary to find out the methods of changing the behaviour and practices of health personnel instead of targeting only people. People have to be provided with experience of care and cure as per their perceptions and not the biomedical perceptions to help change their attitudes and behaviour. Stigma of leprosy is due to the experience of infectivity for hundreds of years which is not curable, and which leads to social exclusion and non-participation as members of groups.

III

Health is an aspect of culture like economic, political, social and religious aspects of life. Culture being an integrated whole, all aspects are interrelated. In any traditional society, the integration is close knit, the static or equilibrium spirit overriding the dynamic nature of culture. Health aspects could be classified into softer or visible aspects such as ill health or disease, which needs treatment. Ideas about prevention and promotion of health and causation of illness are part of hard core of culture, which would fall in the area of belief. The cause-effect relationship about health and illness, therefore, fall in the areas of belief system. The plural treatment therapies, which may be herbal or of synthetic drugs in modern medicine, are guided by results. Magico-religious procedures are also followed in health, more so in tribal society, but not totally absent from the developed societies. There is no vacuum of knowledge or practices in matters of health and medicine in any society. There are professionals in tribal society, generally referred to as Vaidu or Bhagats who have good knowledge of herbal medicine, besides practising magico-religious procedures. Tribal people have faith in them since they are part of the community and are always available.

Development of the concept of welfare state has led the state to assume the responsibility of health of the people. Concepts of health, disease and treatment based on modern biology and modern medicine have created structures and organizations for health care, as we witness them today across the world. Traditional medicine has been reduced to the label of alternative medicine.

Community development programme in 1952, which was a major exercise in planned nation building, had health as one of the prime components. Primary health centres were established as an aspect of community development. Anthropologists were called upon to educate illiterate communities to change their behaviour and attitudes, in order to accept modern methods of agriculture, health care and disease cure systems. Plethora of KAP studies indicated the research role of anthropologists in documenting the baseline information and subsequent change. *India's Changing Villages* by S.C. Dube was one of the first books on community development. Anthropologists like Oscar Levis, David Mendelbaum, George Foster and Morris Oplar came as consultants in this process. The basic paradigm was that people needed to change their behaviour and comply with modern science and technology for their own development and for nation building. Anthropologists thus got the role of social marketing agent of modern science and technology. Traditional medicine was ridiculed as quackery and the concepts of hot and cold, and light and heavy were made fun of.

Anthropology has now found fresh relevance due to primary health care approach in Alma Ata document highlighting importance of intersectoral coordination, which is enshrined in the integrative nature of culture.

In sexuality studies, it has been found that people associated TB with loss of immunity as a result of indulgence in sex. Since HIV and AIDS has a direct relation with the sexual route of transmission, people's ideas on the relationship of sex and immunity get strengthened. The concept of body and metabolism and immunity in Ayurveda and Unani are related to people's expressions. In India, vasectomy could be a major political issue dislodging Prime Minister Indira Gandhi from power. Concern of the adolescent boys about masturbation and loss of semen is summarily rejected in the government training material. People's acceptance of modern diagnostic methods and modern treatment has been misconstrued as acceptance of modern science. Anthropologists do analyse disease causation theories and relief from any form of treatment as hard and soft core of health culture respectively, treatment being the soft and material aspect of culture.

Anthropologists all these years kept quiet under the garb of value neutrality, "a stance which tacitly becomes pro-establishment and anti-change". The strong point of anthropology is its holistic approach and in trying to understand the people's perception, which are necessary for participatory democracy and sustainable development. The universal values about freedom from hunger and disease have to be implemented by adopting an anthropological approach which Nehru labelled as "developing according to the genius of the people". The anthropological approach of looking at health and development issues from their perspectives would facilitate finding solutions including the use of technology which would be acceptable to people, ensuring community involvement in the activities for change.

BIBLIOGRAPHY

Caudill, William. 1953. 'Applied Anthropology in Medicine', in A.L. Kroeber (ed.), *Anthropology Today*. University of Chicago.

Foster and Anderson. 1978. *Medical Anthropology*. New York.

INCAA. 2007. *Human Development, Evolution and Vision: An Abstract Book of Second Congress*. Pune.

Leslie, Charles. 1980. 'Social Science & Medicine: Proceedings of Post-Plenary Session on Medical Anthropology at University of Poona'. December 1978.

McElroy and Townsend. 1979. *Medical Anthropology in Ecological Prospective*.

Mead, Margaret (ed.). 1955. *Cultural Patterns and Technical Change*. UNESCO.

Mutatkar, R.K. 1979. *Society and Leprosy*. Wardha.

——. 1981. *Social and Economic Aspects of Leprosy*, Geneva: WHO.

——. 1984. 'Leprosy: A Challenge to Social Sciences'. Key Note Address, XIIth International Leprosy Congress, New Delhi.

——. 1994. *Social Science & Medicine*. Editorial: The Social Responsibility of Social Scientists.

——. 2003. *Maharashtra Human Development Action Research Study*. Nandurbar, Mumbai: IGIDR.

——. 2007. 'Nation Building and Development: Role of Social Anthropology', in Roy and Danda (eds.), *Methodological Questions in Social Anthropology*. Kolkata.

Mutatkar, R.K. *et al.* 2001. *Curriculum of Anthropology in India*. UGC.

Mutatkar, R.K. *et al.* 2005. *Sexuality and Sexual Behaviour: Social Science Perspective*. University of Pune.

Paul, Bnejamin (ed.). 1955. *Health, Culture and Community: Case studies of public reaction to health programs*. New York.

Planning Commission. 2006. *Report of the Steering Committee on AYUSH in Public Health*. New Delhi.

Pontifical Academy of Sciences. 1984. *Immunology, Epidemiology and Social Aspects of Leprosy*. Vatican.

Spicer, Edward (ed.). 1952. *Human Problems in Technological Change: A Case Book*.

——. 1977. *Ethnic Medicine in the Southwest*. University of Arizona.

LIST OF CONTRIBUTORS

Suresh S. Alizad
Arts, Science and Commerce College, Chopda, Dist. Jalgaon, Maharashtra.

Debopama Bagchi
Dept. of Anthropology, North Bengal University, Darjeeling - 734 430, West Bengal.

H.K. Bhat
Dept. of Anthropology, University of Mysore, Mysore - 570 006, Karnataka.

Subir Biswas
Dept. of Anthropology, North Bengal University, Darjeeling - 734 430, West Bengal.

L.K. Das
Vector Control Research Centre (ICMR), Indira Nagar, Pondicherry - 605 006.

Salil Kumar Dutta
Dept. of Anthropology, Haldia Government College, Haldia, East Midnapore, West Bengal.

C.J. Edwin
Dept. of Anthropology, Panjab University, Chandigarh - 160 014.

Rajan Gaur
Dept. of Anthropology, Panjab University, Chandigarh - 160 014.

S. Gregory
Dept. of Anthropology, Kannur University Campus,
Thalassery - 670 661, Kerala.

K.T. Harichandrakumar
Vector Control Research Centre (ICMR), Indira Nagar,
Pondicherry - 605 006.

C.K. Kantharaju
Dept. of Anthropology, University of Mysore, Mysore - 570 006,
Karnataka.

Sujata Kar
Biological Anthropology Unit, Indian Statistical Institute,
203, B.T. Road, Kolkata - 700 108.

V. Kaul
Anthropological Survey of India, Kaulagarh Road, Dehradun,
Uttaranchal.

Taranpreet Kaur
Dept. of Anthropology, Panjab University, Chandigarh - 160 014.

V.S. Kulkarni
Dept. of Anthropology, University of Poona, Pune - 411 007,
Maharashtra.

Suresh Kumar
Dept. of Anthropology, Panjab University, Chandigarh - 160 014.

A. Krishna Kumari
Vector Control Research Centre (ICMR), Indira Nagar,
Pondicherry - 605 006.

K. Krishnamoorthy
Vector Control Research Centre (ICMR), Indira Nagar,
Pondicherry - 605 006.

Ganesh Chandra Mallick
Dept. of Anthropology, North Bengal University, Darjeeling - 734 430.

S.M. Meera
Dept. of Anthropology, Mysore University, Mysore - 570 006,
Karnataka.

Sailendra K. Mishra
Biological Anthropology Unit, Indian Statistical Institute, 203, B.T. Road, Kolkata - 700 108.

Swati S. Mohanty
Dept. of Anthropology, Sambalpur University, Sambalpur - 768 019, Orissa.

Barun Mukhopadhyay
Biological Anthropology Unit, Indian Statistical Institute, 203, B.T. Road, Kolkata - 700 108.

Shibani Mukhopadhyay
Dept. of Anthropology, Andhra University, Visakhapatnam - 503 003, Andhra Pradesh.

Susmita Mukhopadhyay
Biological Anthropology Unit, Indian Statistical Unit, 203, B.T. Road, Kolkata - 700 108.

R.K. Mutatkar
President, Maharashtra Association of Anthropological Sciences, 64, Anand Park, Aundh, Pune - 411 007, Maharashtra.

Aarti Kaulagekar Nagarkar
School of Health Sciences, University of Poona, Pune - 411 007, Maharashtra.

S. Narahari
Dept. of Anthropology, Andhra University, Visakhapatnam - 503 003, Andhra Pradesh.

Premananda Panda
Dept. of Anthropology, Sambalpur University, Sambalpur - 768 019, Orissa.

S.P. Pani
Vector Control Research Center (ICMR), Indira Nagar, Pondicherry - 605 006.

Purnima Parashar
Dept. of Anthropology, Panjab University, Chandigarh - 160 014.

R.K. Pathak
Dept. of Anthropology, Panjab University, Chandigarh - 160 014.

K. Bhamini Raghavaiah

Dept. of Anthropology, Mysore University, Mysore - 570 006, Karnataka.

K. Rajasekhara Reddy

Dept. of Anthropology, Mysore University, Mysore - 570 006, Karnataka.

Ratnawali

Centre for Social Studies, Surat, Gujarat.

H. Samantaraya

Dept. of Anthropology, Sambalpur University, Sambalpur - 768 019, Orissa.

Sobhanjan Sarkar

Biological Anthropology Unit, Indian Statistical Institute, 203, B.T. Road, Kolkata - 700 108.

Rijumoni Sarma

Dept. of Anthropology, Morigaon College, Assam.

B.V. Sharma

Dept. of Anthropology, University of Hyderabad, Hyderabad - 500 046, Andhra Pradesh.

Teena Singal

Dept. of Anthropology, Panjab University, Chandigarh - 160 014.

C.E. Subramanya

Dept. of Anthropology, University of Mysore, Mysore - 570 006.

P.N. Venugopal

Anthropological Survey of India, Southern Regional Centre, Boghadi, Mysore, Karnataka.